AssocRICS
practical guide to success

Christina Hirst

RICS
the mark of property professionalism worldwide

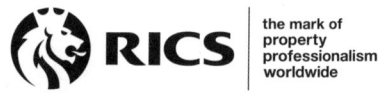

Published by the Royal Institution of Chartered Surveyors (RICS)
Surveyor Court
Westwood Business Park
Coventry CV4 8JE
UK
www.ricsbooks.com

No responsibility for loss or damage caused to any person acting or refraining from action as a result of the material included in this publication can be accepted by the author or RICS.

ISBN 978 1 84219 566 6

© Royal Institution of Chartered Surveyors (RICS) June 2011. Copyright in all or part of this publication rests with RICS, and save by prior consent of RICS, no part or parts shall be reproduced by any means electronic, mechanical, photocopying or otherwise, now known or to be devised.

Typeset in Great Britain by Columns Design XML Ltd, Reading, Berks

Printed in Great Britain by Page Bros, Norwich

Contents

1	Introducing the RICS Associate	1
	The qualification	1
	The benefits	2
	Areas of work	5
	Routes to associate membership	6
	The candidate	8
	Direct entry candidates	8
	Assessment ready candidates	9
	Enrolment ready candidates	10
	Other key people	11
	Associate supporter	11
	Associate proposer	12
	Associate assessors	12
	RICS regional training advisers	12
	Other candidates	13
	Summary	13
2	Competency-based assessment	15
	An explanation	15
	A competency	16
	Competence	19
	Vocational qualifications	19
	What are NVQs or NVQ Diplomas?	21
	How are NVQs achieved?	22

		Benefits of NVQs	23
		Definition of NVQ levels	24
		Who sets the standards?	25
		How does a NVQ link to the Associate assessment?	27
	Apprenticeships		29
	Summary		32
3	The Associate competencies		33
	Overview		33
		Mandatory competencies	33
		Technical competencies	35
	Client care		39
	Communication and negotiation		41
	Conduct rules, ethics and professional practice		44
	Conflict avoidance, management and dispute resolution procedures		44
	Data management		47
	Health and safety		54
	Sustainability		59
	Teamworking		62
	A last word		71
	Summary		71
4	Evidence of competence		73
	Using your existing experience		73
	Submitting your evidence		74
		Choosing your evidence	76
		Uploading your evidence	77
		Linking your evidence to the competencies	78
	Summary		84

5	**Structured training**	86
	Gaining work experience	86
	The Associate supporter's role	86
	Registration	89
	Three-monthly reviews	90
	On completing the minimum experience required	91
	Progressive assessment	91
	Registration	92
	Three-monthly reviews	98
	Actions at the halfway point	103
	Following completion of your minimum period of work experience	103
	Summary	104
6	**Structured development**	105
	The concept of professional development	105
	Structured development for Associate assessment	107
	Stage 1 – Appraisal	108
	Stage 2 – Planning	110
	Stage 3 – Development	111
	Stage 4 – Reflection	113
	Types of structured development activities	115
	Personal	115
	Organised learning	118
	Work-based learning	118
	Completing the structured development record	120
	Summary	124

Contents

7	**The Managed Learning Environment (MLE)**	125
	What are managed learning environments?	125
	The RICS Associate Managed Learning Environment	126
	The MLE in detail	127
	Homepage	127
	Adding evidence	129
	Structured development screens	131
	Viewing your portfolio	134
	Ready for assessment	136
	Summary	138
8	**Professional ethics**	139
	Being a professional	139
	The RICS Rules of Conduct	140
	The candidate's knowledge	143
	Conduct rules, ethics and professional practice	145
	The online ethics module	148
	Some taster questions	148
	Summary	151
9	**The assessment**	153
	Before submitting your portfolio	153
	Checklist	153
	After submitting your portfolio	154
	Checklist	154
	The Associate assessors	155
	Results	156
	Appeals	157
	Quality assurance	158
	Success	159
	Progression from associate membership	160
	Summary	163

10	Conclusion	165
11	Checklists	168
	Assessment submission	168
	Associate supporter action	169
	Ethics module	170
	The 12 RICS ethical standards	172
	The RICS Rules of Conduct for members	174
	The RICS Rules of Conduct for firms	175
	FAQs	177
	Index	185

Introducing the RICS Associate

This chapter introduces you to the RICS' Associate qualification and provides an overview of how you can achieve this. It introduces you to the routes to qualification and to some of the key people who will support you through the process.

The qualification

RICS Associate (AssocRICS) is both an entry-level qualification and a valuable professional qualification in its own right. It provides an opportunity for those with relevant work experience and/or vocational qualifications to gain professional recognition of their skills in the land, property and construction sectors. While full chartered membership of RICS requires degree level entry, the RICS Associate requires evidence of vocational competence. Associate members may be specialists within a particular niche area of practice, may provide technical support for chartered surveyors or may cover broad areas of technical work.

The route to RICS associate membership rigorously assesses competence based on technical skills attained through work experience. You will need to show clear evidence that you are highly competent and that you adhere to RICS Rules of Conduct and ethical standards.

1

Becoming an associate can also provide a stepping stone to achieving chartered status, professional membership (MRICS). Figure 1 below shows the route to the various grades of RICS membership.

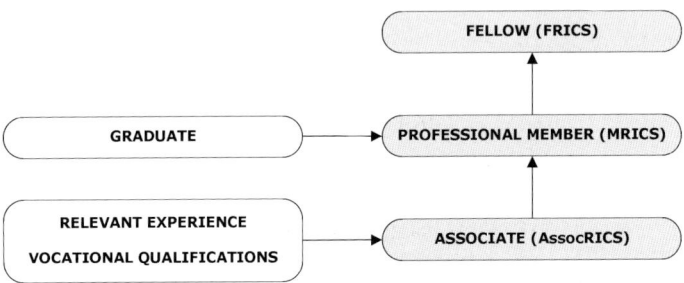

Figure 1 Route to RICS membership

There are different pathways for each specialist area of surveying.

Whatever your academic qualifications, with the right experience and competence you can become an RICS Associate.

The benefits

Career development

The Royal Institution of Chartered Surveyors, commonly known as RICS, is the world's leading professional organisation when it comes to professional standards in land, property and construction.

Membership is only awarded to those who meet and maintain rigorous standards. It requires the formal assessment of professional conduct, ethical values and professional practice. As a result you will hold significant competitive advantage with a CV that stands out from others with many years of experience. You will also be able to progress to chartered status without necessarily needing to attend a traditional university degree course.

In addition, if you have the urge to travel AssocRICS will give you an international passport to employment opportunities across the world.

Status and recognition

RICS is one of the older professional bodies established by Royal Charter in 1868 to set, maintain and regulate standards for the surveying profession and to provide impartial and professional advice to governments, banks and commercial organisations across the world. Today, RICS operates in 146 countries with an extensive network of regional offices located in almost every continent. Over 100,000 property professionals are members of RICS and operate in almost every country in the world. Membership brings with it status and profile that will remain with you throughout your working life.

The RICS designation means that you are part of the world's pre-eminent professional organisation in land, property and construction and shows clear evidence that you have achieved high standards of competence and ethics in your chosen career. The fact that you will demonstrate high ethical values is increasingly important in today's business environment, which is illustrated by the introduction of the Bribery Act 2010 and other ethics based legislation.

Client confidence and employer assurance

Membership is viewed as the definitive mark of surveying expertise. Clients and employers recognise the qualification and are reassured by individuals who can demonstrate that they work to high standards of both technical and professional competence.

Membership also gives you a genuine competitive advantage: increasingly clients are looking to work with organisations with professionally qualified staff and public sector funding and partnerships are often only available for those with quality assured technical competence.

Flexible online qualification process

The user-friendly, online assessment means that you can gain your qualification while in full time employment, without needing to take time off work. It's easy to apply and once you get started you'll have all the help, guidance and support you need. You will need to demonstrate that you meet RICS competency requirements through the presentation of evidence achieved in your day to day job role.

Professional knowledge/information

As an RICS member, you'll be supported by a highly specialist team who offer technical guidance and professional knowledge of the highest calibre. As an Associate you will have access to the world's largest online land, property and construction related library; together with guidance, research and market survey information. You can attend cutting edge events and access continuing professional development (CPD) resources to support your professional development as well as a regular news network, including weekly emails, regional and technical journals.

RICS Matrics

RICS Matrics, the social and professional network for RICS trainees and newly qualified members provides events throughout the UK and offers helpful careers information and advice. Making and maintaining contacts is something you'll need to do throughout your career. RICS Matrics provides a great opportunity to build friendships and professional contacts early on in your career.

Areas of work

RICS has 17 professional groups and within each professional group there are often many specialist areas. RICS associates have a role to play in all of these. RICS professional groups are divided into three key categories:

Property
- Arts and antiques
- Commercial property
- Disputes resolution
- Facilities management
- Machinery and assets
- Management consultancy
- Residential
- Valuation

Built environment
- Building control
- Building surveying
- Project management
- Quantity surveying and construction

Environment and land
- Environment
- Geomatics
- Minerals and waste
- Planning and development
- Rural

AssocRICS: Your practical guide to success

Routes to associate membership

The Associate assessment is open to anyone currently gaining experience in the land, property and construction sectors; there is no requirement for an academic qualification. A typical candidate will be a non-graduate although candidates with non-RICS accredited degrees may also be interested. Whatever your academic background, you will need relevant work experience. Some candidates may need to gain further work experience before taking the Associate assessment or alternatively may choose to undertake additional vocational qualifications to reach the required standard. Each candidate will be assessed on an individual basis.

The Associate qualification is achieved via a simple and easy-to-use online tool so you don't need to take time off during working hours to attend college or university. The qualification is achieved by demonstrating that you meet RICS standards and competencies through evidence of work experience in your day to day role.

There are five steps to achieving the Associate qualification:

Step 1 – Registration

You register online setting out your work experience, existing academic and/or vocational qualifications and/or membership of other professional bodies. This will determine the requirements for your assessment plan. You may be eligible for direct entry and, if so, you will progress directly to step 4.

Step 2 – Assessment plan

Your assessment plan, based on the information you provided during registration, will set out the requirements for you to complete the Associate assessment.

Step 3 – Portfolio of work-based evidence (and structured training if required)

You will be asked to assemble a portfolio of work-based evidence online via the Managed Learning Environment (MLE). The MLE will allow you to:

- record and track your progress against the necessary competencies;
- upload documented work-based evidence to demonstrate your competence;
- upload evidence of relevant qualifications; and
- record structured development and learning outputs.

If you have already satisfied the period of work experience required you will be able to use evidence of your work to date. If not, you will need to undertake a period of structured training to enable you to build up evidence for the required competencies.

Step 4 – Associate assessment

Once you have submitted the evidence required against each competency, it must be signed off by an RICS member who is familiar with your work (typically this would be your line manager). You are then ready to apply for Associate assessment.

A panel of two Associate assessors will review your portfolio online. If you successfully meet the competency requirements, you will pass and qualify as an Associate. Candidates who don't meet all the competency requirements will be referred and advised of the additional evidence required for assessment.

Step 5 – Ethics module

All candidates must successfully complete the RICS online ethics module.

> **TOP TIP**
> Get started as soon as possible – you can then start assembling your portfolio as soon as you have evidence available. Remember you can always change the evidence at a later date.

The candidate

Not all candidates are alike. There are three types of candidate, which can be defined as follows:

- Direct entry – candidates who possess approved NVQ4 or approved professional body membership.

- Assessment ready – candidates who already possess the required amount of work experience.

- Enrolment ready – candidates who are already working or who are about to start in relevant employment and need to complete the required amount of relevant work experience.

Direct entry candidates

If you already have proven competence in your chosen specialism through surveying related qualifications or membership of professional bodies you may be eligible for direct entry to RICS associate membership. If so, you will not be required to build a portfolio of documented work-based evidence to prove your competence and will simply need to submit evidence of your qualifications and to successfully complete the online ethics module. You should contact RICS to obtain a list of the approved qualifications but in essence these will be either an NVQ Level 4 in a relevant subject or a relevant professional qualification (membership of another professional body).

If you are a direct entry candidate see the pages on professional ethics and progression from Associate membership. (If you are a direct entry candidate then the chapters of this book of most relevance to you will be chapter 8 on professional ethics and chapter 9 as regards progression to becoming a chartered surveyor.)

Assessment ready candidates

In order to be ready for assessment you will either need at least four years' relevant experience or will need to satisfy one of the following requirements set out in figure 2 below:

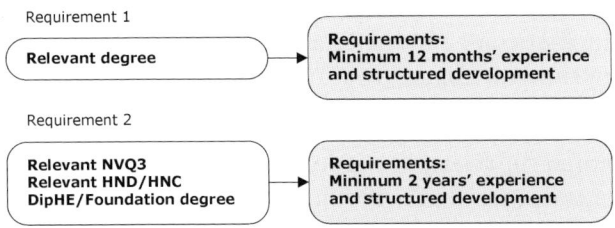

Figure 2 Required experience for assessment

You should contact RICS for guidance on what would constitute a relevant qualification but as a general rule these must relate to property and/or construction and show relevance to your pathway. Examples might include:

- NVQ 3 Surveying, Property and Maintenance
- HNC or HND Construction
- HNC or HND Building Maintenance
- Foundation Degree Valuation
- Foundation Degree Estate Management

> **TOP TIP**
> Check with RICS before you start a course of study to make sure this will be acceptable towards your Associate assessment.

Enrolment ready candidates

To be enrolment ready you need to be in employment that will allow you to gain work experience to achieve the necessary competencies over the required period of time. This could be while you are studying on a relevant qualification. If you do not satisfy the requirements for direct entry and are not yet assessment ready you should enrol as an Associate candidate and then consider the different options to ensure you are ready for Associate assessment. These would include:

- to continue in relevant employment and undertaking work experience that will allow you to achieve the necessary competencies until you have four years' experience;

- to enrol with a university or college to take a HND, HNC, Foundation Degree or other comparable qualification and to continue in relevant employment either while studying or after studying until you have two years' experience. If you study by part-time or distance learning you can collect your evidence while studying. The College of Estate Management's Diploma in Surveying Practice is a distance learning option that you could consider;

- to enrol on a relevant NVQ Level 3. You will need two years' experience as well as your NVQ but these can be gained at the same time (see page 20 for further information about NVQs); or

- to take a surveying related apprenticeship (see page 28 for further information about apprenticeships).

Introducing the RICS Associate

> **TOP TIP**
> If you are able to consider taking an NVQ Level 4 because that would reflect the level of your work you might consider taking an approved NVQ Level 4 as this will allow you to take the direct entry route to associate membership (see page 8).

When considering the most appropriate option you should take into account the following:

- your employer's views and requirements for you;
- your style of learning;
- whether you would like face to face tutor support;
- whether your employer will allow you time to attend a college course;
- your family circumstances and the time you will have available for study;
- whether you would like a formal academic qualification;
- whether you would like a formal vocational qualification; and
- the amount of experience you already have.

> **TOP TIP**
> The option you choose to take you to the point of being ready for Associate assessment should be the one that you feel best suits your circumstances and learning style.

Other key people

There will be a number of key people with whom you will work to achieve your Associate qualification.

Associate supporter

Unless you are ready immediately for Associate assessment, you will benefit from having an Associate supporter. This will most likely be your line manager, or a suitable person in your organisation who is able to give you guidance and verify the evidence submitted is substantially your own work. Your Associate supporter should also be involved in planning and structuring your period of work experience to ensure that this will enable you to meet the necessary competencies.

Associate proposer

Your Associate proposer must be an associate (AssocRICS) of four years' standing, a member (MRICS) or fellow (FRICS) of RICS. Your Associate proposer will endorse your application by signing a declaration form at Associate assessment to confirm that you are a fit and proper person to practise.

Associate assessors

There will be two assessors involved in assessing your evidence. They will be:

- an Associate assessor – an associate (AssocRICS), member (MRICS) or fellow (FRICS) who assesses your submitted evidence via the Managed Learning Environment (MLE) and decides whether you have met the requirements of your pathway.
- a lead Associate assessor – as above, but with the extra responsibility of writing the feedback for referred candidates and managing the contact between the two assessors before a decision is reached.

RICS regional training advisers

RICS employs regional training advisers (RTAs) for each region in the UK. Details may be found on the RICS website, at www.rics.org

RTAs can advise on all aspects of Associate assessment and importantly can provide advice and support for your employer, your Associate supporter and your Associate proposer. They can also work with your employer to help develop an Associate structured training agreement.

Other candidates

Remember that other candidates can help you with your Associate assessment. There is nothing better than sharing ideas and approaches with others doing the same thing. [The RICS isurv Associate channel will give you an excellent online opportunity to network with others using discussion forum discussions and Facebook or LinkedIn links.] For face to face networking and opportunities for discussion, RICS Matrics is a great way to meet other candidates. For details of RICS Matrics, the branch of RICS for young surveyors, trainees and Associate candidates take a look at the RICS website (www.rics.org/matrics).

Summary

- RICS Associate is available for anyone who is able to achieve the necessary competencies by gaining relevant work experience.
- There are different pathways to RICS Associate that reflect the different surveying specialisms.
- Your route to RICS Associate will depend upon your starting point, i.e. qualifications and/or experience.
- There are three key types of Associate candidate: direct entry, assessment ready and enrolment ready.
- There are a number of key people with whom you will work, in particular your Associate supporter and Associate proposer.

- Don't overlook the opportunities to discuss your Associate assessment with other candidates.
- It is essential to read the general guidance on the RICS Associate website at www.ricsassociate.org

Competency-based assessment

In this chapter we introduce you to the concept of competency based assessment around which the Associate assessment is based. We discuss what is meant by competence and consider formal qualifications such as NVQs that assess vocational competencies. We also consider how the NVQ can link with your Associate assessment.

An explanation

The Associate assessment determines your 'competence to practice', in other words your competence to carry out the work of an RICS associate. To be 'competent' is to have the skill or knowledge to carry out a task or function successfully – this ability can vary from being merely able, to being expert in a particular sphere of activity.

The assessment requires you to achieve various competencies relevant to your pathway. The competencies are a mix of technical, professional and personal skills and come in three categories:

- mandatory competencies;
- core competencies; and
- optional competencies.

The core and optional competencies will be different for each Associate pathway. The Associate assessment requires you to submit evidence that demonstrates that you meet with the knowledge requirements for each competency of your chosen pathway. In addition, you will need to demonstrate that you can apply this knowledge in practice.

You will also need to show that you have been developing the range of knowledge and skills required in the mandatory competencies, which comprise a mix of professional practice, interpersonal, business and management skills that are considered common to, and necessary for, all surveyors.

A competency

A competency is a statement of the skills or abilities required to perform a specific task or function. It is based upon attitudes and behaviours, as well as skills and knowledge. For the Associate assessment the requirements for each pathway are set out in the relevant pathway guides and on the RICS Associate website – the competency definitions are also reproduced on isurv Associate.

In the broader context, the use of competency-based training and assessment is a global concept. Many employers use competency-based human resource management that involves the development of organisational competency frameworks, around which they develop recruitment and training policies. Interviews and staff personal development reviews are all related to the competencies.

Competency-based assessment is a fundamentally different approach to traditional academic courses. It is not a set of examinations: it is the basis for certification of competency and it is carried out as a process in order to collect evidence about the performance and knowledge of a person with respect to a set competency standard.

Traditional assessment systems usually have some or all of the following characteristics:
- assessment is associated to a course or programme;
- the course is divided into subjects that are assessed independently of one another;
- parts but not all of the programme are included in final examinations;
- passing criteria are based on marking scales;
- questions can be ignored;
- it is done within limited periods of time; and
- statistical comparisons are used.

On the other hand, competency-based assessment is defined as a process with the following steps:
- setting of objectives;
- collection of evidence;
- comparison of evidence with objectives; and
- opinion formation (competent or not yet competent).

Some of the characteristics of competency-based assessment are that:
- it is based on standards that describe the expected level of individual competence;
- standards include criteria that provide details of what is considered to be good practice;
- the assessment is individual – there is no comparison among individuals;
- it provides a judgment for the assessed person: competent or not yet competent;
- it is done, preferably, in or from real working situations;

- it does not take a predetermined period of time – it is a process rather than a particular moment;
- it is not subject to the completion of a specific training action;
- it includes the recognition of acquired competencies as a result of experience;
- it is a tool for the orientation of subsequent learning or training and as such it plays an important role in the development of skills and abilities; and
- it is the basis for the certification of labour competence.

For example, the process can be described as follows:

- The candidate to be assessed is introduced to the certifying body (RICS);
- this body carries out a pre-diagnosis of competencies;
- the candidate is referred to an assessment centre (online Managed Learning Environment);
- an assessor is appointed;
- an assessment plan is drawn up;
- a portfolio of evidence is applied and integrated;
- an assessment judgment is issued;
- there is a positive verdict regarding certification;
- the certification is issued.

Based on this vocational profile, the contents for training and development plans can be drawn up by establishing the necessary theoretical and practical professional knowledge for a competent performance of units. In order to do this, the unit of competency is taken as the grounds for the analysis and the following questions are answered.

- What does the individual need to know in order to establish the theoretical knowledge?
- What does the individual need to know how to do in order to establish the practical knowledge?
- How does the individual need to act in order to show the required attitudes and behaviours?

Competence

There are multiple and diverse concepts of competence. A widely accepted concept defines it as the effective ability to perform a fully identified vocational activity successfully. Competence is not the possibility of success at a job; it is a real and proved ability.

A good categorisation of competence is one that distinguishes between three approaches. The first approach regards competence as the ability to carry out tasks; the second one concentrates on personal attributes (attitudes, abilities) and the third one, called 'holistic', is a combination of the two previous approaches.

Vocational qualifications

Vocational qualifications offer a formal way of assessing competence and competencies in the UK. They can be used as an entry to Associate assessment. Obtaining a relevant level 3 vocational qualification will half the period of required work experience from four years (for someone with no vocational qualification) to two years. This section provides some general background and advice regarding these qualifications.

The Qualifications and Credit Framework (QCF) is a new framework that contains vocational (or work-related) qualifications, available in England, Wales and Northern Ireland. These qualifications are made up of units that are worth credits. You can study units at your own pace and build these up to full qualifications of different sizes over time. Every qualification and unit on the QCF has a credit value, showing how long it takes to complete. One credit is equivalent to 10 hours. This can include time spent learning in a variety of ways – not necessarily being taught by someone. There are three different sizes of qualification, worth different numbers of credits; larger qualifications will take longer to complete than smaller ones:

- **Award**

    An award is the smallest type of qualification on the QCF. It is worth between one and 12 credits. This means it takes between 10 and 120 hours to complete.

- **Certificate**

    A certificate is worth between 13 and 36 credits. It takes between 130 and 360 hours to complete.

- **Diploma**

    A diploma is worth 37 credits or more, so takes at least 370 hours to complete.

The units are transferrable to other qualifications and you can complete a unit in one part of the UK, and then combine it with other units that you take somewhere else in the UK.

Each type of qualification on the QCF also has a level between Entry level and level 8, to show how difficult it is. The QCF levels are the same as the levels on the National Qualification Framework, which lists other types of qualification. For example, GCSEs are at levels 1 and 2, A levels are at level 3, and a PhD is at level 8.

The title of every qualification on the QCF contains the following information:

- what the qualification is about;

- the size of the qualification (award, certificate or diploma) – how long it takes; and
- the level of the qualification (entry level to level 8) – how difficult it is.

This means you can see at a glance how much time you'll need to complete the qualification, and how difficult it is likely to be compared with other qualifications.

You need to obtain a qualification at at least level 3 for this to count towards Associate assessment.

Qualifications that existed before the launch of the QCF in 2010 are being changed into units and moved on to the QCF. Some of these qualifications are still called NVQs, BTECs, City and Guilds, OCRs, HNCs and HNDs but also refer to their QCF type (diploma, certificate or award in their title). Other pre-existing qualifications are being totally renamed. Most NVQs relevant to surveying will become NVQ Diplomas or just Diplomas.

What are NVQs or NVQ Diplomas?

The central feature of NVQs is the National Occupational Standards (NOS) on which they are based. NOS are statements of performance standards that describe what competent people in a particular occupation are expected to be able to do. They cover all the main aspects of an occupation, including current best practice, the ability to adapt to future requirements and the knowledge and understanding that underpins competent performance.

NVQs and their new replacement, NVQ Dipolmas, are work-related, competency-based qualifications. They reflect the skills and knowledge needed to do a job effectively, and show that an individual is competent in the area of work the NVQ represents.

NVQs do not generally have to be completed within a specified time. They can be undertaken by full-time employees or through college with a work placement or part-time job that enables them to develop the appropriate skills. There are no age limits and no special entry requirements.

How are NVQs achieved?

NVQs are achieved through assessment and training. Assessment of the NVQ is normally through on-the-job observation and questioning and also the presentation of evidence of work done. Evidence is either provided by performing your duties or showing examples of your work; the aim is to prove you have the competence to meet the NVQ standards. The assessor tests your underpinning knowledge, understanding and work-based performance to determine whether you can demonstrate competence in the workplace; the assessor will 'sign-off' units when you are ready.

When you start a NVQ the assessor will usually help you to:

- identify what you can do already;
- agree on the standard and level of NVQ you are aiming for;
- analyse what you need to learn; and
- choose and agree on activities that will allow you to learn what you need.

At this point you may need to agree with your employer to do slightly different work to gain the level of competence you need.

You will compare your performance with the standards as you progress in order to determine what you have achieved, how much you still need to achieve and how you should go about achieving it, until you are assessed as competent for a unit of a whole NVQ.

NVQs can be used if you already have skills and want to increase them, but it is also for those who are starting from the beginning. As the system is flexible, new ways of learning can be used immediately.

Benefits of NVQs

For you

NVQs can help you prepare for work or help your career development and are achieved through the demonstration of skills. Obtaining a NVQ shows you can do a job to national standards and have the up to date skills that employers are looking for. In terms of Associate assessment, the evidence you use for your NVQ can also be used towards your Associate assessment where it is relevant to the competencies for your chosen pathway.

With NVQs the practice is equally as important as the theory – they show you what you can do as well as validating what you know. You don't have to sit through hours, days, weeks and in some cases even years of a learning programme – and there are no exams to sit at the end. NVQs are about putting learning into practice.

There are virtually no limits with NVQs – no time limits (although you should look to do it within a defined timeframe to ensure the currency of your competence), no age limits and certainly no limits to the opportunities they create! Generally a level NVQ or NVQ Diploma will take between one and two years to obtain assuming that you are in relevant employment.

For your employer

Employers play a vital role in their contribution towards the development of standards for NVQs so they have been developed to ensure they meet employers' needs.

NVQs can help employers improve productivity and competitiveness and ensure you have the skills and knowledge to meet business needs. Many major companies use NVQs and find them valuable tools for both the development of employees and their business. Significant benefits have been reported by employers in terms of better productivity levels, an improvement in employee motivation, lower staff turnover, better staff-manager relations, and improved staff recruitment. They also provide the opportunity to benchmark standards and provide training.

Definition of NVQ levels

NVQs are organised into five levels, based on the level of competence required. The following definitions provide a general guide to the progression from level to level and the relationship between them.

Level	Description
Level 1	Competence that involves the application of knowledge in the performance of a range of varied work activities, most of which are routine and predictable.
Level 2	Competence that involves the application of knowledge in a significant range of varied work activities, performed in a variety of contexts. Some of these activities are complex or non-routine and there is some individual responsibility or autonomy. Collaboration with others, perhaps through membership of a work group or team, is often a requirement.

Level	Description
Level 3	Competence that involves the application of knowledge in a broad range of varied work activities performed in a wide variety of contexts, most of which are complex and non-routine. There is considerable responsibility and autonomy and control or guidance of others is often required.
Level 4	Competence that involves the application of knowledge in a broad range of complex, technical or professional work activities performed in a variety of contexts and with a substantial degree of personal responsibility and autonomy. Responsibility for the work of others and the allocation of resources is often present.
Level 5	Competence that involves the application of a range of fundamental principles across a wide and often unpredictable variety of contexts. Very substantial personal autonomy and often significant responsibility for the work of others and for the allocation of substantial resources features strongly, as do personal accountabilities for analysis, diagnosis, design, planning, execution and evaluation.

Who sets the standards?

There are a number of organisations involved in the process of developing, delivering, awarding and preserving the quality of NVQs:

- *Sector bodies* identify, define and update employment-based standards of competence for agreed occupations.

- *Awarding bodies* design assessment and quality assurance systems, and gain sector bodies endorsement.
- *Qualification and Curriculum Framework*, which is regulated by Ofqual, accredit the qualification.
- *Awarding bodies* approve assessment centres to offer NVQs, and implement and assure quality of the NVQs.
- *Assessment centres* offer and assess NVQs.

Sector bodies

National Occupational Standards upon which NVQs are based are set and designed by the relevant sector body. This is either a Sector Skills Council – the government-licensed body representative of their sector's skills needs – or, where they do not exist, the body recognised by the sector as representing their skills needs. In the surveying world, most National Occupational Standards are set by either Construction Skills or Asset Skills, both of which are government-licensed Sector Skills Councils.

The relevant sector body determines the content, structure and assessment strategy for NVQs. Achievement is certifiable at unit level, with mutual recognition between awarding bodies. NVQs are accredited against NVQ criteria by QCF.

You can find all accredited NVQs on the National Database of Accredited Qualifications (www.accreditedqualifications.org.uk)

Qualifications and Curriculum Framework

The QCF will ensure that NVQ qualifications meet particular criteria and are broadly comparable across different sectors. Ofqual oversees the regulation of the QCF.

Competency-based assessment

Awarding bodies

Awarding bodies have a dual role. With sector bodies, they are jointly responsible for the assessment methods of NVQs based on the assessment strategy of the sector bodies, and they are also responsible for the implementation of individual NVQs. They approve centres that wish to offer assessment for NVQs.

Awarding bodies monitor the assessment process and award NVQs and unit certificates. They undertake external verification to ensure that candidates are being assessed fairly and consistently across all centres.

Assessment centres and training providers

Many candidates pursuing the NVQ route to qualifications will gain their qualification at work or through a programme provided by a further education college or some other training provider.

Many local further education college or local training providers provide surveying related NVQs or NVQ Diplomas.

Scottish Qualifications Authority (SQA)

In Scotland, the NVQ equivalent is the Scottish Vocational Qualification (SVQ). The system of vocational education and training differs from that of the rest of Britain. The Scottish Qualifications Authority (SQA) accredits all SVQs.

How does a NVQ link to the Associate assessment?

NVQ Level 4

If you obtain a Level 4 NVQ or NVQ Dipolma that has been approved by RICS you will gain direct entry to the online ethics module without the need for compilation of further evidence on the Managed Learning Environment.

NVQ Level 3

If you obtain a Level 3 NVQ or NVQ Dipolma you may be assessment ready if you have also completed at least two years' work experience (see figure 2 on page 9). You will still need to upload evidence to the Managed Learning Environment but you will be able to make use of the same evidence that you used for your NVQ because many of the Associate competencies are very similar to the NVQ unit requirements for a relevant NVQ.

> **TOP TIP**
> Taking a NVQ gives you a formal vocational qualification and you can use much of the same evidence for your Associate assessment. Holding a relevant NVQ Level 3 can half the period of work experience required from four years to two. Holding an approved NVQ Level 4 can exempt you from the required period of work experience and give you direct entry to the online ethics module.

The following is an example of how you might use evidence from your NVQ for your Associate assessment:

The Level 3 NVQ Dipolma in Surveying Property and Maintenance has been designed for built environment practitioners working in a variety of fields.

- In a maintenance occupation you are likely to be involved in the assessment of the condition of property with a view to planning, implementing and monitoring its maintenance, including procuring and monitoring works.

- In a quantity surveying occupation you are likely to be involved in measuring and drafting bills of quantities and contributing to project procurement and managing project cost control.

- In a general practice occupation you are likely to be involved in the assessment of the condition of property with a view to contributing to the processing of property agreements, including acquisitions, disposals and also property management.

- In a building surveying occupation you are likely to be involved in the assessment of the condition of property with a view to monitoring works, including procurement and organising work programmes.

- In a valuation occupation you are likely to be involved in the assessment of the value of a property.

If you were taking the quantity surveying and construction pathway to Associate qualification you would need to achieve the competency entitled 'Quantification and costing of construction works'. One of the NVQ units you would take would be 'Measure and draft bills of quantities'. Clearly, in order to show your competence in this NVQ unit you would present a bill of quantities that you had prepared; this same bill could be presented as evidence for your Associate assessment competency – there is no need to find a different example.

Apprenticeships

As employees, apprentices earn a salary and work alongside experienced staff to gain job-specific skills. Off the job, usually on a day-release basis, apprentices receive training to work towards nationally recognised qualifications. Anyone living in England, over 16 and not in full-time education can apply for an apprenticeship. Apprenticeships can take between one and four years to complete depending on the level of apprenticeship, the apprentices' ability and the industry sector.

There are three levels of Apprenticeship available for those aged 16 and over:

1 – Immediate Apprenticeships (equivalent to five good GCSE passes)

- Apprentices work towards work-based learning qualifications such as an NVQ Level 2, Key Skills and, in some cases, a relevant knowledge-based qualification such as a BTEC.
- These provide the skills you need for your chosen career and allow entry to an Advanced Apprenticeship.

2 – Advanced Apprenticeships (equivalent to two A-level passes)

- Advanced apprentices work towards work-based learning qualifications such as NVQ Level 3, Key Skills and, in most cases, a relevant knowledge-based certificate such as a BTEC.
- To start this programme you should ideally have five GCSEs (grade C or above) or have completed an Apprenticeship.
- The advanced apprenticeship provides you with qualifications that count towards your Associate assessment.

3 – Higher Apprenticeships

- Higher Apprenticeships work towards work-based learning qualifications such as NVQ Level 4 and, in some cases, a relevant knowledge-based qualification such as a Foundation Degree.
- A higher apprenticeship can provide you with the opportunity to become an RICS Associate by direct entry.

Apprenticeship programmes are based on frameworks developed by Sector Skills Councils in partnership with employers. The actual apprenticeship itself is then delivered by a training provider or a college.

To use an apprenticeship to become an RICS Associate you would need to take at least an Advanced Apprenticeship. There are a number of relevant apprenticeships available, including:

- Property, Surveying and Maintenance Apprenticeship;
- Facilities Management Apprenticeship; and
- Housing Apprenticeship.

All of the above include an NVQ Level 3 or Level 3 Diploma that would, in my opinion, be relevant for your Associate assessment. They also include a Technical Certificate, which is a qualification (sometimes a HNC) that will help you to develop the necessary knowledge and understanding for both your NVQ and Associate assessment.

For further advice you should contact your local college, ideally one that offers construction courses, to discuss this further. Alternatively you should contact the relevant Sector Skills Council:

Construction Skills at www.cskills.org.uk

Asset Skills at www.assetskills.org.uk

If you are aged 16 to 24 and are finding it difficult to access your surveying career, perhaps because of financial circumstances, because you have been unable to find work or because you have other reasons why you might need additional support, you should consider contacting the Chartered Surveyors Training Trust (CSTT). The CSTT is a charity that supports young people who face a barrier to becoming a surveyor. The CSTT offers an surveying apprenticeship programme in association with RICS that meets the requirements for RICS Associate and beyond. When you finish your apprenticeship the Trust can also offer support for degree level study. The Trust can be contacted at www.cstt.org.uk

Summary

- The Associate assessment assesses your 'competence to practice'.
- Associate assessment is competency-based.
- There are three types of competency for each Associate pathway: mandatory, core and optional.
- Mandatory competencies are the same for all pathways but should be set in a relevant context for each pathway.
- NVQs are formal qualifications that assess vocational competence.
- NVQs can be used towards your Associate assessment.
- An apprenticeship can provide you with a programme to become an RICS Associate.
- You can use the same evidence that you used for your NVQ for your Associate assessment, assuming this is relevant to the Associate competency.

3 The Associate competencies

In this chapter we provide some advice and ideas as to the knowledge and understanding you should develop for each of the mandatory competencies.

Overview

There are three types of Associate competency: mandatory, core and optional. The core and optional competencies are known together as technical competencies and differ for each pathway. The mandatory competencies are the same for all Associate pathways although the context within which they are set will differ for each pathway.

Mandatory competencies

The mandatory competencies are:

- Client care;
- Communication and negotiation;
- Conduct, rules, ethics and professional practice;
- Conflict avoidance, management and dispute resolution procedures;
- Data management;
- Health and safety;

- Sustainability; and
- Teamworking.

All of the above competencies are defined on the RICS Associate website and these definitions are also given in the discussion on each competency in this chapter.

The mandatory competencies are one of the most important parts of the whole Associate assessment. The skills and abilities they encourage and test underpin all professional and technical aspects of working as a surveyor, and are vital for further advancement in the profession. They also reflect skills that are transferrable from other occupations or work experience that you may have on starting the Associate assessment process.

This chapter will provide some further explanation of the requirements for each mandatory competency and practical advice on how to achieve the necessary level of competence.

The advice is not intended to be all-encompassing or definitive in any way – each candidate, each pathway, and each firm will alter the ingredients of each competency slightly. It is certainly not possible to say, 'follow this, and you will pass'. What follows are simply suggestions and pointers as to how the requirements of the competencies might be met.

In particular, you should note that the recommended reading is not exhaustive – it is, in the main, a selection of some useful texts and websites. Each pathway will have its own specialist authors and texts, and you should talk to your Associate supporter asking for their views and recommendations based on their experience, which should be more tailored to your pathway.

I would also encourage you to use the RICS online service – *isurv* (www.isurv.com). This service, mixing expert commentary with official RICS guidance, covers a

The Associate competencies

huge range of surveying matters, and can be used for professional development purposes.

You can use *isurv Associate* to help you: the mandatory competencies are listed with links to relevant content on other *isurv* channels, thus allowing you to develop knowledge of these competency areas.

For each of the mandatory competencies discussed you should refer to the official RICS guidance for the detailed definitions and for further advice; the following sections simply provide further practical guidance to support these definitions.

Technical competencies

The technical competencies are those that are at the heart of your pathway and have been set and agreed by qualified surveyors from that pathway. Many of these are compulsory but there may be others that are optional. The technical competencies for each pathway are linked to below:

- Quantity surveying and construction;
- Residential survey and valuation;
- Residential estate agency.

For example in the construction and quantity surveying pathway the core competencies are:

- construction technology and environmental services;
- contract practice;
- procurement and tendering;
- project financial control and reporting; and
- quantification and costing of construction works.

The optional competencies are either:

- commercial management of construction (which will be relevant if you are working in a contracting or commercial environment); or

- design economics and cost planning (which you will be relevant if you are working in a consulting environment within either the public or the private sector).

All Associate assessment pathways will require you to meet around six technical competencies.

Where optional competencies are offered, you should choose these carefully and ensure they match the work that you have or will be doing. You should work closely with your Associate supporter, if you choose to have one, in choosing these competencies. You will demonstrate your attainment of these competencies by submitting evidence of your work in the Managed Learning Environment.

An example of a technical competency is set out below:

Competency – Contract practice (from the quantity surveying and construction pathway)

Competency description: This competency covers the various forms of contract used in the construction industry. Candidates should have an awareness of all of the main standard forms of contract and an understanding of contract law, legislation and the specific forms of contract they have used.

Requirements of competency: Demonstrate knowledge and understanding of the various forms of contract used in the construction industry and/or your area of business. Apply your knowledge of the use of the various standard forms of contract at project level, including the implications and obligations that apply to the parties to the contract.

Examples of likely knowledge:

- Basic contract law and legislation
- Contract documentation
- The various standard forms of contract and sub-contract
- When different forms would be used
- Basic contractual mechanisms and procedures at various stages of the contract
- Third party rights including relevant legislation and the use of collateral warranties

Examples of likely skills and experience:

- Producing contract documentation
- Carrying out the contractual mechanisms and procedures relevant to the financial management of aspects of the project such as change procedures, valuations and final accounts
- Understanding general contractual provisions such as insurances, retention, bonds, liquidated and ascertained damages, early possession and other common contractual mechanisms

You will be required to submit documentation as evidence for each competency. In this example evidence would be required of your involvement in preparing the following:

- preliminaries, employer's requirements or works information document;
- completion of contract details;
- interim valuation (including statement of retention and valuation recommendation); and
- statement of final account.

The evidence submitted must clearly show your involvement with the piece of work and may specify that you also show how you dealt with certain matters. In this example, evidence would be required of your involvement in:

- queries in preparing preliminaries, employer's requirements or works information document;
- insurance provisions;
- ascertained damages;
- variations;
- assessing preliminary items;
- adjustment and agreement of valuations/final accounts;
- unfixed materials on/off site; and
- taxation.

RICS practice statements and guidance

Whichever Associate pathway you take you will need to be familiar with the relevant RICS practice statements and guidance relating to your area of work or particular tasks. Remember that it is mandatory for RICS members to follow the requirements of practice statements and it is recommended that members follow guidance notes. Examples include:

For valuation: RICS Appraisal and Valuation Standards – Red Book

For residential estate agency: RICS Estate Agency Standards and guidance – Blue Book

For measurement: RICS Code of Measuring Practice

A full list of RICS official material is available on *isurv*.

Developing knowledge for your technical competencies

A key component of technical competence is knowledge upon which practical application is then built. You may be taking an academic course to develop this knowledge, for example, a HNC or a Foundation Degree. You should, however, use your structured development to support your attainment of the technical competencies (see Chapter 6). You can use *isurv Associate* to help you: the technical competencies are listed for each pathway with links to relevant content on other *isurv* channels, thus allowing you to develop knowledge of your competency areas.

> **TOP TIP**
> You will need to use your structured development to help you to meet the technical competencies for your pathway; *isurv* is an excellent resource to assist with this.

The mandatory competencies in detail

The following sections consider each mandatory competency in detail and provides practical guidance and some useful tips.

Client care

The requirements of this competency are that you should be able to demonstrate a knowledge and understanding of the principle and practice of client care, including:

- the concept of identifying all types of client and the behaviours that are appropriate to establish good relationships; and
- the systems and procedures for managing the client care process – including complaints, data collection and client needs analysis.

Practical guidance

Approaches to client care will vary from business to business, depending on the nature of the work, the degree of client interface and the type of organisation. At one extreme – residential valuation, sales and letting – there is a high degree of customer interface, and the skill of managing and influencing clients is vital, with the business relying solely on fee income for survival. In areas of central or local government, by contrast, the link is not so direct and the dependency on fee income from transactions not as great.

However, one key concept is fundamental to all business: the client is sovereign. Notice that this is not the same as saying, 'the customer is always right' – in surveying matters this is not always the case!

You must be able to understand the link between customer care and duty of care. If a client wishes to do something that would be impractical, or impossible, or doomed to certain failure, the surveyor owes the client a duty of care to inform them of that. To give the best customer care, you must therefore have a good understanding of each client's needs.

This is one of those competencies where you must be able to 'step back' from a situation, to analyse what it is that you have learnt about customer care and duty of care in any particular circumstance. To assist your understanding of this subject, you should also undertake some structured reading and training.

This competency is very closely related to professional ethics and practice.

> **TOP TIP**
> Think about who your clients are. This may be simple if you are in the private sector but if you are in the public or corporate sector, your client may be another department, a board or an individual within your organisation.

Communication and negotiation

The requirements of this competency are that you demonstrate a knowledge and understanding of effective oral, written, graphic and presentation skills, including methods and techniques appropriate to specific situations.

Practical guidance – communication skills

Oral communication is used in a wide range of surveying situations and circumstances: at meetings, in negotiations, when managing people, when making presentations, in tenders, and so on. In my experience, there is some basic best practice for all situations. This can then be tailored to meet the requirements of specific situations. The list of specific situations is of course huge, and training will also be wide and varied.

Communication skills can be improved by additional training. Many organisations run courses in this area – these may include, for example, assertiveness training courses. However, for most candidates, the best approach is to be coached by senior people in your particular area, and to put your developing skills continually into practice continually.

On a general level, any course or training programme should cover the nature and purposes of oral communication – addressing the different approaches to be taken in different situations, and the techniques that can be used to communicate effectively.

In addition, there are many texts on the subject, allowing training to be complemented by structured reading. A search on the RICS Books website for 'communication skills' will identify a number of excellent publications.

Written and graphic communication will cover a wide range of situations and encompass a variety of skills. It will include the use of emails, letters and reports, for all of which an essential component is being able to write good, clear paragraphs.

You must be able to understand the various media in which written communications can be presented, and more importantly, the skills involved in doing so, with regard to the target audience, the length, style and layout of the communication, the message you wish to convey, and the structure of the communication. Graphic communication covers sketch notes, drawings in plans and designs linked to the construction process and similar (if these are relevant to your chosen pathway).

It is probably easiest to assess understanding with reference to actual written work – your own and other people's – in a variety of mediums. Consider why a particular communication fails or succeeds, how it could be improved, and what aspects make it successful in a particular area. These days there are so many different types of communication medium that it is becoming ever more important to ensure the method you use is appropriate. Methods include:

- letters;
- emails;
- reports;
- telephone;
- meetings;
- telephone conferences;
- web conferences;
- video conferences;
- online social networking;
- blogs;
- text messages; and

- web messaging,

to name but a few.

If you are experiencing difficulties in this area, you could attend a course (internal or external) on written communication. There are also a lot of useful books and texts on the subject, again available through RICS Books. Nobody would expect you to write a perfect client report the first time you tried, but with a full understanding of the principles and purposes of the report, and some more practice, you will be much better equipped to do so.

Practical guidance – negotiation skills

This competency overlaps somewhat with the conflict avoidance competency.

To fulfil the requirements of this particular competency, check that you understand what lies behind successful negotiations: the preparation of evidence; an understanding of the various approaches to negotiations; a knowledge of where and how parameters are set; a knowledge of what each side wishes to get from the negotiations, and from any future relationship; and so on.

If possible, ask your manager or another colleague to involve you in negotiations carried out by your firm. This is perhaps the best way of helping you to gain an understanding of principles and skills.

> **TOP TIP**
> Think about who you communicate with and what methods of communication you use and why. Do you always use the most appropriate method of communication?

Conduct rules, ethics and professional practice

This competency is achieved through the online ethics module undertaken at the end of the Associate assessment. See chapter 10 for further information regarding the RICS Rules of Conduct and the online ethics module.

Conflict avoidance, management and dispute resolution procedures

This competency requires knowledge and understanding of the techniques for conflict avoidance, conflict management and dispute resolution procedures, including for example adjudication and arbitration appropriate to your Associate pathway.

Practical guidance

The 'ingredients' of this competency will vary greatly between the various pathways. In commercial practice, for example, landlord and tenant matters will be fairly common, while in construction, this competency will be present everyday in managing building contracts. In basic terms, and across all pathways, it is important that you understand how to conduct negotiations, and also the various options available should negotiations break down, working through mediation and conciliation, adjudication, arbitration, independent expert determination, and, finally, litigation.

You may be encouraged to sit in on negotiations at your firm from an early stage in your career. Also you will benefit from some formal training on this and other aspects of dispute resolution, covering the preparation of evidence, case law, approaches and tactics. It is reasonably likely that by the time you reach the end of your assessment you will have had practical experience of running your own negotiations, or participating in other dispute resolution procedures, and will thus be able to discuss this. As part of your training plan ensure that you make steady progress towards this end.

There are also many texts available on this subject. To get started, the following RICS guidance notes and practice statements are useful documents:

- *Surveyors acting as arbitrators and as independent experts in commercial property rent reviews*;
- *Surveyors acting as adjudicators in the construction industry*;
- *Surveyors acting as expert witnesses*; and
- *Surveyors acting as advocates*.

All of these publications are available in hardcopy from RICS Books, at www.ricsbooks.com RICS members can download the publications as a PDF from www.rics.org

In addition, don't forget CPD-type lectures or training that may be available within your firm, or from external providers.

Specific issues to be aware of are:

In the UK

PACT

Professional Arbitration on Court Terms (PACT) is a scheme offered by RICS and the Law Society for the resolution of lease renewal disputes. PACT is aimed at unopposed lease renewals under the *Landlord and Tenant Act* 1954 and offers the opportunity for disputes to be resolved without the necessity of going to court.

The scheme provides the opportunity for landlords and tenants to have the terms and rent payable under their new lease decided by a surveyor or solicitor acting as either an arbitrator or independent expert. RICS or the Law Society can make the appointments. The professionals appointed are experienced specialists who have been specifically trained under the PACT scheme.

The objective of the scheme is to increase the effectiveness and flexibility of the legal system and to give a greater choice to both landlords and tenants and to their advisers. The scheme is in line with Lord Woolf's reports on Access to Justice, which advocated that litigation should be viewed as a last resort, and also the provisions of the *Arbitration Act* 1996 and the Civil Procedure Rules for litigation.

PACT is an attempt to streamline the lease renewal process, making it quicker, cheaper and more efficient than going to court.

Mediation

Mediation is the name given to a confidential process whereby parties to a dispute invite a neutral individual to facilitate negotiations between them with a view to achieving a resolution of their dispute.

Arbitration

An arbitration is a legal proceeding under the Arbitration Acts and the arbitrator reaches a decision on the basis of evidence put before him or her sometimes at a formal hearing. The arbitrator can call for discovery of documents and interprets the evidence. The arbitrator's decision is enforceable as if it were a judgment of the court. Although the arbitrator is not liable for negligence the court can set the judgment aside on the grounds of misconduct.

Independent expert

An independent expert is appointed jointly by the two parties to carry out a normal valuation and to give their own expert opinion on the matter to be decided. The expert may have regard to evidence submitted or may have a hearing and adopt what they consider to be the most appropriate procedure. The expert's decision is not enforceable directly by the courts and they are liable for action for negligence.

The RICS Disputes Resolution Professional Group website provides some really good guidance in this area. See www.rics.org/disputeresolutionpg

For this competency you should also ensure that you are aware of the RICS requirements regarding complaints handling.

> **TOP TIP**
> This competency is not just about formal methods of dispute resolution – it is just as much about your personal skills in managing conflicts. Think about conflicts you have dealt with within your own office, for example.

Data management

This is a mandatory competency.

This competency involves demonstrating knowledge and understanding of the sources of information and data applicable to your area of practice, including the methodologies and techniques most appropriate to collect, collate and store data.

Practical guidance

Again, this competency will vary greatly between pathways. It is important to think of it in relation to your specific pathway, and against the backdrop of your day-to-day work and the particular IT developments in your area.

In the quantity surveying and construction pathway sources of data may be previous contracts or cost guides and price books. Various commercial packages are also available to price contracts and bills of quantities.

For all pathways, the important thing is for you to be able to understand the use of data in your day-to-day work – how this is gathered and put to use, and what the best methods of collection, collation and storage are. You should be able to step back mentally from your work, to explain what data you use, how you find it and how it is manipulated. You should also be aware of the implications of data protection legislation and how this will affect the use that can be made of data you may hold.

Try to use the competency to broaden and develop your understanding of wider data issues and developments in the profession. See this competency as a subject in itself and carry out some structured reading.

Within your work you will use many different sources of information, including:

- colleagues;
- documents;
- the internet;
- data;
- books;
- journals;
- companies;
- government departments;
- files; and many more.

It is worthwhile just thinking over your recent work and considering what information you have used and where it came from. Think also of other information that your organisation may have provided to others. Who needed this and why?

When you have information you need to consider what you are able to do with it. This may not be as straightforward as you might think because of laws that deal with the handling of information.

In the UK

You need to familiarise yourself with the following legislation.

The *Data Protection Act* 1998 gives individuals the right to know what information is held about them. It provides a framework to ensure that personal information is handled properly.

The Act works in two ways. Firstly, it states that anyone who processes personal information must comply with eight principles, which make sure that personal information is:

- fairly and lawfully processed;
- processed for limited purposes;
- adequate, relevant and not excessive;
- accurate and up to date;
- not kept for longer than is necessary;
- processed in line with your rights;
- secure; and
- not transferred to other countries without adequate protection.

The second area covered by the Act provides individuals with important rights, including the right to find out what personal information is held on computer and most paper records.

Should an individual or organisation feel they are being denied access to personal information that they are entitled to, or feel their information has not been handled according to the eight principles, they can contact the Information Commissioner's Office (ICO) for help. Complaints are usually dealt with informally, but if this is not possible, enforcement action can be taken.

The Data Protection Act does not guarantee personal privacy at all costs, but aims to strike a balance between the rights of individuals and the sometimes competing interests of those with legitimate reasons for using personal information. It applies to some paper records as well as computer records.

This short checklist will help you understand how to comply with the Data Protection Act. Being able to answer 'yes' to every question does not guarantee compliance, and you may need more advice in particular areas, but it should mean that you are heading in the right direction.

- Do I really need this information about an individual? Do I know what I'm going to use it for?

- Do the people whose information I hold know that I've got it, and are they likely to understand what it will be used for?

- If I'm asked to pass on personal information, would the people about whom I hold information expect me to do this?

- Am I satisfied the information is being held securely, whether it's on paper or on computer? And what about my website? Is it secure?

- Is access to personal information limited to those with a strict need to know?

- Am I sure the personal information is accurate and up to date?
- Do I delete or destroy personal information as soon as I have no more need for it?
- Do I need to arrange for the Information Commissioner to be notified?

The *Freedom of Information Act* 2000 came into force at the beginning of 2005. It deals with access to official information, while other regulations deal with environmental information.

The Act provides individuals or organisations with the right to request information held by a public authority. They can do this by letter or email. The public authority must tell the applicant whether it holds the information, and must normally supply it within 20 working days, in the format requested. However, the public authority does not have to confirm or deny the existence of the information or provide it if an exemption applies, the request is vexatious or similar to a previous request, or if the cost of compliance exceeds an appropriate limit. If exemption applies, but is qualified, this means that the public authority must decide whether the public interest in using the exemption outweighs the public interest in releasing the information.

If an applicant is unhappy with a refusal to disclose information, they can complain to the ICO, after first exhausting any internal review procedure. The ICO will investigate the case and either uphold the authority's use of an exemption or decide that the information must be disclosed.

Information must also be published through the public authority's publication scheme. This must be approved by the ICO, and is a commitment by a public authority to make certain information available, and a guide on how to obtain it.

The Act applies to all information, not just information filed since the Act came into force.

In Scotland the relevant legislation is the *Freedom of Information (Scotland) Act* 2002.

Environmental Information Regulations

The *Environmental Information Regulations* give members of the public the right to access environmental information held by public authorities. The request can be made by letter, email, telephone or in person.

The regulations apply to most public authorities, but they can also apply to any organisation or person carrying out a public administration function, and any organisation or person under the control of a public authority who has environmental responsibilities. This can include some private companies or public private partnerships, for example companies involved in energy, water, waste and transport.

Environmental information is divided into the following six main areas:

1 The state of the elements of the environment, such as air, water, soil, land, fauna (including human beings).

2 Emissions and discharges, noise, energy, radiation, waste and other such substances.

3 Measures and activities such as policies, plans, and agreements affecting or likely to affect the state of the elements of the environment.

4 Reports, cost-benefit and economic analyses.

5 The state of human health and safety, and contamination of the food chain.

6 Cultural sites and built structures (to the extent they may be affected by the state of the elements of the environment).

If a public authority receives a request for information on any of the areas mentioned above, they are legally obliged to provide it, usually within 20 working days. There are a number of exceptions to this rule – for example, if the information is likely to prejudice national security – and if this is the case, the public authority must explain why the exception applies.

Public authorities must provide advice and assistance to applicants when necessary. A reasonable charge may be made for environmental information. No charge may be made for environmental information held in registers or lists or for viewing the information at the public authority's premises. There is no 'appropriate limit' to the cost of providing environmental information.

If you are dissatisfied with the way the public authority has dealt with your request, you may put your comments to the public authority and it must then reconsider its decision. If you are still dissatisfied after the reconsideration procedure, you can make a complaint to the ICO and they will investigate the case independently and act on its conclusions. The applicant can appeal the ICO's decision to the Information Tribunal.

Scotland has its own Scottish Environmental Information Regulations.

Issues to consider

Some issues to ensure you are aware of in this area of data management legislation would be:

- What the Information Commissioner's Office is.
- Examples of when you have had to be mindful of data management legislation.
- What information you may need to request under the Freedom of Information legislation.
- The key principles of the relevant legislation.

Health and safety

The requirement of this competency is to demonstrate a knowledge and understanding of the principles and responsibilities with regard to health and safety imposed by law and codes of practice and other regulations relating to health and safety appropriate to your area of practice.

Attainment of the Construction Skills Certification Scheme Card will significantly contribute to this competency. Further details regarding the Card Schemes can be found at www.cscs.uk.com. The website explains what level of health and safety test would be relevant for you and the card that you should hold. Generally for surveyors working on site you will need a manager level health and safety test and a platinum card. For an assistant surveyor you may only need the gold card.

Practical guidance

This competency covers all aspects of a surveyor's working life. It is about ensuring that the surveyor's entire working life is conducted as safely as possible with as little risk to health as possible, and that the same is true for all of those around the surveyor.

It is easy to think of ways in which health and safety issues relate to, say, work on a construction site, but perhaps less so for more office-based work. However, the same basic philosophies underpin all work carried out in any environment. In off-site jobs, the issues encompass such things as your managers knowing where you are and what you are doing at all times, and, should you leave the office, when you will return and who you are meeting. There are also numerous health and safety issues relating to the use of equipment, in offices as well as all other locations, and on keeping employees healthy and safe.

Owing to the importance of health and safety, most firms and organisations conduct formal training and instruction on the relevant issues. Ensure that you attend this and that you can explain the reasons behind any requirements imposed by the firm.

You must be able to demonstrate knowledge of the health and safety legislation and codes of practice that apply to your area of work and the country in which you work. These will differ from one pathway to another. You must be aware of relevant legislation and your evidence should allow you to demonstrate how you have developed this knowledge.

In the UK

Relevant legislation would include the *Health and Safety at Work Act* 1974, the *Construction (Design and Management) Regulations* 2007 and the *Control of Asbestos Regulations* 2006. You should also know and understand the role of the Health and Safety Executive (HSE) which is responsible for health and safety regulation in England and Wales.

HSE owns a significant amount of primary and secondary legislation. The primary legislation comprises the Acts of Parliament, including the *Health and Safety at Work etc Act* 1974. The secondary legislation is made up of Statutory Instruments (SIs), often referred to as regulations. It is enforced by HSE and local authorities (LAs). The HSE and LAs work locally, regionally and nationally to common objectives and standards.

The Health and Safety Executive (HSE) provides numerous free leaflets on its website, at www.hse.gov.uk, including lists of its current publications. You should visit the site and select some useful reading. The website also provides interesting statistics.

The HSE publication *Health and Safety in Great Britain* provides an excellent summary of the legal framework for health and safety in the UK and could be a useful source of structured development for you in this area.

Risk assessments

A risk assessment is a very important element in assessing health and safety risks and you should know and understand how to undertake a risk assessment of your workplace. I would recommend that you undertake a risk assessment of your own workplace: this could be a useful source of evidence for your Associate assessment and could form part of your structured development.

In the UK

The HSE *Five Steps to Risk Assessment* document is a very useful guide to performing a risk assessment.

Personal safety

Consider your own personal safety and how you ensure this when at work. Health and safety can also include issues of personal safety and you should consider how you ensure your own safety when out on site, at meetings or travelling for work. You should consider the following actions:

- Take a fully charged mobile phone with you.
- Carry a highly audible personal alarm.
- Do not lock doors behind you.
- Plan your escape route, ideally in two directions.
- Implement a call back system with your office – for example, telephone to say that you are about to start the inspection and that you will phone back at a pre-agreed time to confirm you are safe. Make sure your office knows your phone number as well just in case they need to call you.
- Make your daily work schedule available to others so that they can trace your steps.

- Be very careful about inspecting a roof void and safely position your ladder.
- Park your car close by and ensure you cannot be boxed in. Keep you car keys with you.
- Do not try to move any heavy equipment.
- Make sure the person who greets you is the person you expected to meet. If not, and you are unsure of their status, make an excuse and delay the survey.
- If met by someone under the age of 16 you should delay the survey until such time as an adult is present.
- If the person you meet appears angry, stay calm and leave if they become abusive or you feel afraid.
- Undertake your survey alone (without the occupier going round with you) and explain that you need to do this in order to concentrate.
- If you interrupt someone – for example you walk in on someone in bed – leave the room and apologise from the other side of the door.
- Keep your equipment close by you in case you did need to leave.
- Follow your instinct. If you feel threatened or uncomfortable make an excuse and leave and return with a colleague at another time.

Safety of property

You should always take sensible precautions to protect your personal property, whether you are at home or work.

Electronic gadgets, whether mobile phones, cameras or laptops, are attractive targets for thieves; don't leave them lying around on site or in the office or on view when you leave your car.

In the UK

You can register any consumer product that has a serial number for free on the UK National Property Register (www.immobilise.com). It's quick and convenient, provides you with a register of property and, in case of loss or theft, can easily be identified and returned to you if found.

Remember:

- Backup your phone numbers and photos in a separate location as a safety precaution.
- Take extra care with clients personal details that maybe stored on electronic equipment.
- Keep all banking details secure, your own and those of clients.
- Use a cross-cut shredder to destroy all personal data if it doesn't need to be kept.
- Keep your personal documents and those of clients, safe from electronic theft.

Find out more about how you can protect yourself at www.identity-theft.org.uk.

Personal property of others

As a surveyor you will be visiting properties, homes or sites owned by others and there are a number of issues for you to consider.

For example:

- Always remove dirty shoes when you enter a property. To make a good impression you could carry some indoor shoes in your equipment bag.
- Take extra care when using tape measures and ladders in confined spaces. It's very easy to knock over an ornament or mark a wall.

- Make sure you leave the space as you found it – put back any items that you move, clear up debris you may have caused and if you damage anything, advise the occupier that your employer will be in contact to arrange suitable compensation.
- Don't use the occupier's personal property without asking, whether it's a broom, a cloth or the downstairs loo!
- Don't read any personal correspondence, private documentation or messages – even if it relates to your task.
- If you see any sexually compromising material, ignore it and resist the urge to make a comment.
- If you feel that there is anything that could be compromise your health or safety make an excuse, leave and discuss it with your employer.

> **TOP TIP**
> Remember health and safety can be just that but also can be personal safety, safety of your personal property and the safety of property of others.

> **TOP TIP**
> If you visit construction sites or are likely to, consider gaining your Construction Skills Certification Scheme Card. You will need this when out on site and it can also be used as evidence towards this competency.

Sustainability

For RICS the principle of sustainability seeks to balance economic, environmental and social objectives at global, national and local levels, in order to meet the needs of today without compromising the ability of future generations to meet their needs.

AssocRICS: Your practical guide to success

This competency requires a knowledge and understanding of why and how sustainability seeks to balance economic, environmental and social objectives at local, national and global levels, in the context of land, property and the built environment.

Practical guidance

All RICS associates should have at least a good, basic understanding of environmental issues, which range from groundwater pollution and contaminated land, to control of pollution in the air we breathe, as well as, very importantly, energy and climate change. Environmental issues affect building design, construction use and management, development and redevelopment, and regeneration and town planning. Issues such as global warming, dwindling national resources and atmospheric pollutants are top priorities with many government and influential bodies, such as the EU and the World Trade Organisation (WTO).

You should carry out some general reading in newspapers and professional journals on environmental issues. Other useful sources of information and advice are the RICS guidance notes and information papers:

- *Carbon management of real estate*;
- *Contamination and environmental issues – their implications for property professionals*;
- *Renewable energy*; and
- *Sustainability and the RICS property lifecyle.*

All of these publications are available in hardcopy from RICS Books, at www.ricsbooks.com. RICS members can download the publications as a PDF from www.rics.org

Surveying Sustainability: a short guide for the property professional, produced in partnership between RICS,

Forum for the Future and Gleeds, and the sustainability pages on the RICS website (www.rics.org/sustainability) are also useful.

In addition, of course, you should maintain an awareness of environmental issues while at work. You should be aware of any government initiatives, laws or EU regulations affecting your particular area of work. The Deparment of Energy and Climate Change (DECC) website is a useful resource (www.decc.gov.uk). On a more local level, be aware of any internal office environmental policies (recycling of paper, for example), and be able to explain the purposes of these. You may like to test your knowledge and understanding in this area by considering how you would express the firm's 'green credentials', should this be requested, for example, in an invitation to tender. Once more, this is a case of stepping back from day-to-day work, to consider the environmental factors that underlie and overarch such work.

Some useful websites for this competency are:

- Department for Energy and Climate Change: www.decc.gov.uk
- Department for Environment Food and Rural Affairs: www.defra.gov.uk
- Department for Communities and Local Government: www.communities.gov.uk
- Carbon Trust: www.carbontrust.co.uk
- Energy Saving Trust: www.energysavingtrust.org.uk

> **TOP TIP**
> Policy and legislation regarding sustainability are constantly evolving. You should ensure you keep up to date by reading daily newspapers or news sites regularly and property or construction journals.

Teamworking

For this competency you will need to demonstrate a knowledge and understanding of the principles, behaviour and dynamics of teamworking.

Practical guidance

This competency involves understanding why people work in teams, and some of the basic principles underlying teamworking. In practice, you will rarely not work with a team. Evidence of working in a team will be easy to come by; however, the requirement is to *understand the principles* behind this. You should therefore consider a situation in which you have witnessed or experienced teamworking, and be able to explain how that team worked, concentrating on the roles each member adopted and the success or otherwise of this.

Your understanding can be complemented and extended by some reading on the subject. A leading text on this subject is *Management teams: Why they succeed or fail*, by R. M. Belbin, (Butterworth-Heinemann, Oxford, 2003).

Background

In the past thirty years or so, teamworking has grown in importance. Until relatively recently, roles at work were well-defined. In the traditional manufacturing industry, for example, there was strict division of responsibilities and most job titles conveyed exactly what people did. However, with advances in technology and education, employers began to place a growing emphasis on versatility, leading to an increasing interest in teamworking at all levels. The gradual replacement of traditional hierarchical forms with flatter organisational structures, in which employees are expected to fill a variety of roles, has also played a part in the rise of the team.

What is a team?

There are numerous definitions, but for our purposes a team is simply defined as a limited number of people who have shared objectives at work and who cooperate, on a permanent or temporary basis, to achieve those objectives in a way that allows each individual to make a distinctive contribution.

Types of team

There are many types of teams. The following is not a comprehensive list, and there are many other classifications that you may find.

- Production and service teams – construction is a good example of where you will find production teams. They have a relatively long life-span, providing an ongoing product or service to customers or the organisation.

- Project and development teams – including research and product development teams. They are dedicated to a particular objective, and have limited life-spans and a clear set of short-term objectives. They are often cross-functional, with members selected for the contribution their expertise can make.

- Advice and involvement teams – with the aim of improving, for example, working conditions or quality. Members will not devote a great deal of time to them, and once they have achieved their objectives they should be disbanded.

- Crews – such as airline crews, who may be formed from people who have rarely worked together but through prior training clearly understand their respective roles.

- Action and negotiation teams – such as surgical and legal teams, consist of people who tend to work together regularly. They have well-developed processes and clear objectives.

- Virtual teams – who work in separate buildings and who may even be in different countries. Such teams may also fit into one of the above categories, such as project and development. They may need to communicate by telephone, email and tele-conferencing rather than face-to-face. Managing them is particularly difficult, not least because remote working can exacerbate misunderstandings.

- Self-managed teams – where much decision-making is devolved from line managers to team members. (Also known as semi-autonomous or fully autonomous teams according to the degree of self-management.) Again, such teams may also fit into one of the above categories. It has been suggested that the positive benefits of teamworking may be largely associated with such teams, rather than teams in general. Fully or semi-autonomous teams tend to have higher than average levels of labour productivity, a lower rate of voluntary resignations, lower levels of employee dismissals, and a better than average employee relations climate.

Teams may consist of people from the same employer, but equally they may include professionals from different employers – for example on a construction project.

Benefits of teamworking

Organisations have introduced teamworking for the following reasons, among others:

- to improve productivity;
- to improve quality of products or services;
- to improve customer focus;
- to speed the spread of ideas;

- to respond to opportunities and threats and to fast-changing environments;
- to increase employee motivation; and
- to introduce multi-skilling and employee flexibility.

There can be benefits for employees too. The most commonly-quoted outcomes are greater job satisfaction and motivation, and improved learning. However, the introduction of teamworking needs skilful management and resources devoted to it, or initiatives may fail.

Stages of team development

The main stages of team development are generally considered to be as follows:

- Forming (or undeveloped) – when people are working as individuals rather than a team.
- Storming – teams need to pass through a stage of conflict if they are to achieve their potential. The team becomes more aggressive, both internally and in relation to outside groups, rules and requirements.
- Norming (or consolidating) – the team is beginning to achieve its potential, effectively applying the resource it has to the tasks it has, using a process it has developed itself.
- Performing – when the team is characterised by openness and flexibility. It challenges itself constantly but without emotionally charged conflict, and places a high priority on the development of other team members.
- Mourning – when the team disbands.

While this is a useful theoretical model, it should not be seen as unvarying. For example, a mature performing team may revert to an earlier stage if something happens, perhaps the loss of a key member or a threatening change in the organisation. Alternatively, a team in which the members know each other well may perform almost from the start.

Characteristics of effective teams

An effective team has the following characteristics:

- a common sense of purpose;
- a clear understanding of the team's objectives;
- resources to achieve those objectives;
- mutual respect among team members, both as individuals and for the contribution each makes to the team's performance;
- a valuing of members' strengths and respecting their weaknesses;
- mutual trust;
- a willingness to share knowledge and expertise;
- a willingness to speak openly;
- a range of skills among team members to deal effectively with all its tasks;
- a range of personal styles for the various roles needed to carry out the team's tasks.

Team role theories and team selection

There are two requirements in selecting team members: the team should include a range of the necessary technical and specialist skills, and there should be a variety of personal styles among members to fill the different roles that are involved in successful teamwork. The pioneering work on team roles or types was carried out by Dr Meredith Belbin in the 1970s. He lists nine team roles:

- Plant – creative, imaginative, unorthodox; solves difficult problems.
- Resource investigator – extrovert, enthusiastic, exploratory; explores opportunities; develops contacts.
- Coordinator – mature, confident, a good chairperson; clarifies goals; promotes decision making.
- Shaper – dynamic, challenging; has drive and courage to overcome obstacles.
- Monitor evaluator – sober, strategic, discerning; sees all options.
- Teamworker – cooperative, mild, perceptive, diplomatic; listens, builds, averts friction.
- Implementer – disciplined, reliable, conservative; turns ideas into practical action.
- Completer – painstaking, conscientious, anxious; searches out errors and omissions, delivers on time.
- Specialist – single-minded, self-starting, dedicated; provides knowledge and skill in rare supply.

There has been some criticism of Belbin's work on the grounds that individuals rarely fit neatly into these categories – most fit into more than one, and arguably the best teamworkers will adapt their behaviour to fill different roles as circumstances require. However, knowing that one tends to fit a certain profile arguably has value in understanding one's own and others' strengths and weaknesses.

There are now many other psychological tests that result in different team type classifications. Team selection is not an exact science and instinct should be taken into account when putting a team together. A mix of types is necessary, as is a mix of skills.

Teams can include senior and junior people (for the latter, team membership may also be a development opportunity) and someone relatively junior may be a team leader. What is most important is the team's mix of skills and types.

Team size

Most commentators suggest that between five and eight people is the ideal size for teams. Teams need to be large enough to incorporate the appropriate range of expertise and representation of interests, but not so large that people's participation, and hence their interest, is limited.

Team leadership

Leadership is vital for successful teams.

There is no one recipe for successful team leadership. Like other team members, team leaders have their own personal styles, which they need to understand and work within:

- Some people, by instinct, will be directive – they will want to tell people what to do. Those with directive tendencies will need to temper their approach to avoid causing resentment; otherwise, other team members may ask 'If he knows all the answers, why are we involved?'

- Others will be democratic, and ask questions to gain commitment and get people on board, even if they themselves have clear ideas about how things should be done. Leaders with democratic tendencies will need to be aware that there is a danger of drift and lack of direction if there is too much debate.

- Some leaders will be more involved, while others will let team members get on by themselves.

Whatever their personal styles, leaders should:

- listen to team members;
- question them to understand their points of view; and
- be responsive to feedback.

In this way they act as facilitators or coaches to get the most out of team members, and to encourage learning and creativity. The roles that leaders play, and hence the ways in which they behave, may differ at different stages of team development – for example, helping to overcome conflict in the early stages and setting tasks at a later stage. It can be argued that successful team leaders need a high degree of emotional intelligence.

In some circumstances, leadership may rotate – for example, different individuals may take the lead at different stages of a project for which a team is responsible. Some semi-autonomous or fully autonomous teams appoint their own leaders.

Team training and learning

Team-building training is often necessary to assist the move from working in a traditional hierarchy to being part of a team, and in circumstances where team members have not worked together previously and may not even know each other. Such training may consist of exercises carried out jointly under a facilitator, sometimes outdoors to enable people to get to know each other and to work together, allowing them to understand each other's strengths and weaknesses; there are many team building activities adopted by organisations.

Communications, knowledge-sharing and problem-solving may often be on the agenda, but the areas covered will depend on the nature and role of the team, so it is impossible to generalise.

Social events may also be used to allow team members to know each other. Separate training may take place for team leaders. As projects develop, there may be additional training that emerges from the team's needs; an important role of the team leader is to act as a coach or facilitator to encourage learning.

Team reward

It is a criticism of traditional staff appraisal systems that they give insufficient weight to individuals' contributions to teams, but this is tending to change. A number of organisations have introduced team pay systems, aimed at encouraging group endeavour rather than individual performance. Research has found that such schemes are less important for success than management style, culture and the working environment. Where this is introduced, it should be done with great care, and the complementary impact of non-financial reward should always be acknowledged.

And finally on teamworking

Teams come in many forms and exist for many purposes. Teamworking is desirable in many circumstances and, properly managed, can contribute to improved organisational performance, while improving individuals' job satisfaction and helping to empower them. But not all teams succeed. Inadequate terms of reference, poor selection of team members, inadequate resources, the wrong mix of personality types and skills, the wrong size, inadequate training and poor leadership are among the reasons why teams fail.

> **TOP TIP**
> While there are academic theories of teamworking some of which I have described here, you should consider the teams you work in and with. What do you think make these successful?

A last word on mandatory competencies

Hopefully you now have a better idea of what the mandatory competencies entail, and of how to help you achieve them. Remember that the philosophies behind the mandatory competencies, and the business skills inherent in them, will be encountered in every aspect of your working life. It is for this reason that they are mandatory!

Remember also that you will be able to use your structured development to help you to achieve the mandatory competencies and also any relevant qualification you may take towards becoming an RICS Associate. The RICS isurv channels and, in particular the RICS Associate channel gives an excellent opportunity to access relevant material to help you develop your knowledge for these competency areas.

Summary

- There are eight mandatory competencies.
- The mandatory competencies are one of the most important parts of the whole Associate assessment. The skills and abilities they encourage and test underpin all professional and technical aspects of working as a surveyor, and are vital for further advancement in the profession.
- You should consider all the mandatory competencies in the context of your pathway and your own work environment.
- Technical competencies may be core or optional. You will need to choose optional competencies carefully and in consultation with your Associate supporter.
- Technical competencies are different for each pathway.

- You will need to demonstrate achievement of the competencies by linking your evidence to them (see chapter 4).
- You will need to undertake a wide range of reading to develop your knowledge within the mandatory competencies. This can form part of your structured development (see chapter 6).
- Academic theory is useful but practical knowledge and application is of more importance.

4 Evidence of competence

Using your existing experience

When you register as an Associate candidate you will be asked about any previous experience you have. You will need to make a judgment as to how much of your experience is relevant to the pathway for which you are applying.

Your previous experience does not need to relate to all the technical competencies for your pathway but must relate to some. Where you have not yet satisfied a competency or are unable to provide sufficient evidence for it, you will need to 'top up' your experience. You should discuss this with your Associate supporter.

You will be asked how many years of experience you have. This should be given as a whole number of years and should be rounded up (or down) accordingly. You should analyse your experience carefully, mapping your activities against the competencies before confirming how much experience you have. For example, if you have worked as a self-employed building contractor for the last ten years, you should consider how much of your working time was spent on issues related to the technical competencies for your pathway. You may find that you only have two years of relevant experience – the result of spending an occasional few hours on a particular task throughout your total ten years' experience. Therefore, your 'relevant experience' should be logged as two years.

You should discuss this with your Associate supporter in order to gain a second opinion and I would recommend that you set out a table with each competency identifying work you have done under each heading and the total experience you have of doing that item of work.

> **TOP TIP**
> Don't fool yourself here – it is better to err on the side of caution and take a little time to plug gaps in your experience than to come forward for assessment without robust evidence of your competencies.

The outcome of your registration will be to advise you of the number of years' experience you need to be ready for Associate assessment. The number given is the total number of years and you will be able to use your previous experience. Therefore, if as in the example above you have two years experience and the result of your registration is that you need four years of experience then you will need two additional years that are relevant to the competencies. This may of course be longer in total if your experience is not largely relevant to the pathway competencies.

Submitting your evidence

You will need to submit evidence to be assessed for each competency. The evidence you submit must be a real output from work that you have undertaken and must clearly show your own personal involvement. Any evidence you submit must have been produced within the last four years and at least one piece of evidence per competency must have been produced within the 12 months immediately prior to your Associate assessment– candidates with at least 10 years' experience can benefit from a variation to this requirement.

RICS recognise that candidates with at least 10 years' experience may be working in a narrower but more senior capacity and therefore may be unable to provide evidence that has been produced during the 12 months prior to Associate Assessment. Candidates in this situation can choose up to three technical competencies for which this requirement does not apply. Evidence must still have been produced within the last four years and must be brought up to date in the commentary, for example by making reference to work candidates have supervised.

You must make sure that your evidence clearly shows the dates when the work was undertaken either by the specific inclusion of a date or the date of a report or by clarifying this within your commentary (see page 79). You should choose evidence that illustrates the breadth and depth of your knowledge and skills and therefore you should use as much variety in your types of evidence as possible.

Examples of evidence you may submit include (depending upon your pathway):

- letters, emails or other correspondence you have written;
- reports you have prepared (even if these were ultimately signed off by someone else);
- drawings, valuations, or other relevant material you have prepared;
- notes of activities you have been involved with;
- work programmes you have prepared;
- notes or minutes of meetings clearly showing your participation;
- risk assessments;
- audit reports you either prepared or that clearly show your involvement;

- recommendations made to clients or other stakeholders; and
- option appraisals.

Choosing your evidence

You should choose evidence that, where relevant, demonstrates your participation in dealing with issues that were not necessarily completely straightforward and involved you in questioning issues, coordinating work with others and dealing with more complex issues. It is unlikely that evidence relating to the first time you have done something will be the evidence you ultimately submit as the level of your involvement and the complexity of the issues you deal with will expand the more often you undertake a particular action. You should discuss your choices of evidence regularly with your Associate supporter.

> **TOP TIP**
> You can upload evidence to the Managed Learning Environment and change this or replace it with other evidence at any time until you finally submit your portfolio for assessment.

Where you have been involved in only a small element of work relating to a particular report or if only a small part of a report or other document is relevant to the competencies, you should only submit an extract and not the whole document.

Generally evidence will be in the form of documents produced through your work but it is also acceptable to use evidence from an academic course if this meets the time requirements (that it was produced within the last four years). However, any work from an academic course that is submitted must relate to your job and not just the general work of a surveyor from your pathway. For example, in many courses tutors will set an assignment that asks you to use something from your work. This would be a valid source of evidence. As a general rule though RICS advises that no more than half of your evidence overall should be from coursework produced for an academic qualification. If you are studying for an NVQ or other vocational qualification you will be required to submit work-based evidence for this and therefore you should be able to use much of the same evidence from your NVQ for your Associate assessment.

> **TOP TIP**
> Whatever evidence you submit do make sure that this is well written, and that there are no spelling or grammatical errors. Bear in mind that the simple production of evidence for a technical competency is a great way for the Associate assessors to assess the mandatory competency of communication!

You should check that any evidence submitted is technically accurate. I would urge you to ask your Associate supporter to check this before you decide to include it within your portfolio but bear in mind that the evidence submitted should be your work – you should only be gaining advice from your supporter and proposer.

Uploading your evidence

When you have decided to submit a piece of evidence you must produce it as PDF file. This is the only format accepted by the MLE (see chapter 7). Follow the instructions in the MLE guide to upload the files to the site – the guide is available on the MLE and within the pathway guides. If you wish to submit a document that you have only in hard copy, it must be scanned for uploading. You must give each document a unique title when you upload it for identification.

If you do not have access to scanning facilities either at work or home you will able to use public services such as your local library or internet café. Make sure you keep a copy and keep a relevant backup as you would in your normal practice.

You do not need to submit the documents in any particular order, and at any time before you submit your evidence for assessment you can change your mind about a document. If you upload a document that you actually wish to replace with a better example of your work at a later stage you will be able to do this. You can change your evidence at any time up until the point you submit the documentation for assessment.

Linking your evidence to the competencies

You must submit four pieces of evidence for each technical competency. Do not be concerned if one piece on its own does not demonstrate the whole range and depth required. Choose evidence that taken together builds up a picture, reflecting different aspects of your work. For each competency, the Associate assessors will be considering all four pieces together and looking at the bigger picture they present.

Each piece of evidence can be linked in the MLE to one technical competency only – so choose the one it mainly reflects. It will then count as one of your four pieces for that competency, and the MLE will 'count down' until you have lodged the required number for all your competencies.

If you have produced a piece of evidence that you think demonstrates more than one of your technical competencies you should prepare another version of this evidence for the second technical competency and upload it as a separate document. It must be given a separate title and you will write a separate 300-word commentary for it. Remember that the Associate assessors will want to see the breadth of your work experience; you should, therefore, try to use as many different examples as you can rather than reusing a single piece of work several times.

Note: Each piece of evidence can be linked to multiple mandatory competencies.

For each piece of evidence, you must submit a 300-word commentary. This serves three purposes:

- to demonstrate how you have understood the requirements of the technical competency and explain how the piece of evidence demonstrates that you have achieved it – in effect, you are explaining why you chose this particular piece;
- to demonstrate your understanding of the mandatory competencies and explain how they are reflected in the work that provided the piece of evidence (for example, did you have to work with other team members or demonstrate communication skills); and
- to set out the process you followed to complete the activity covered by your evidence.

The commentary is important. It shows how you have reflected on what is required, and on your own work, and builds up a picture of what your work involves and how you go about it. You must be concise, as you have a strict word limit. There is no prescribed form for a commentary but you may find it helpful to use these headings:

- **How the competency is demonstrated:** The requirements for each competency should guide you in this; the requirements are set out in section 1.1 of each Associate pathway guide.
- **Wider skills:** Other than the main technical competency, what else does this evidence show about

your work? Look particularly at the definitions of the mandatory competencies and explain how this piece of evidence demonstrates that you have achieved one or more of them.

- **Background:** Describe the work that led to the piece of evidence – where, when, how? Who was working with you? How much supervision did you receive? Is the activity part of your everyday role? How much experience do you have in the activity?

Example commentary

The following is an example of a commentary for the quantification and costing of construction works competency from the quantity surveying and construction pathway.

The details of the competency are as follows:

> **Description:** This competency covers the measurement and definition of construction works in order to value and control costs. It covers the candidate's understanding and involvement with the various methods of quantifying and pricing construction works used throughout a project.
> **Requirements:** Demonstrate knowledge and understanding of the principles of quantification and costing of construction works as a basis for the financial management of contracts. Apply your knowledge to the quantification and costing of construction works, including the use of appropriate standard methods of measurement and forms of cost analysis. Carry out measurement and costing of works at all stages of the construction process.

Evidence of competence

Examples of likely knowledge, skills and experience:

Knowledge
- The quantification of construction works (including both measurement and definition);
- The various standard methods of measurement;
- The costing of construction works; and
- The measurement of buildings and structures to agreed standards.

Examples
- Quantifying construction works at the various stages of a project;
- Producing pricing documents such as bills of quantities, schedules of activities/works, schedules of rates or contract sum analyses;
- Carrying out the costing of construction works by methods such as tendered rates, quotations or dayworks; and
- construction works by methods.

Evidence: Evidence should demonstrate involvement with the preparation of the following:

1 Manual or computerised take offs/measurement or remeasurement of site works;

2 Pricing documents such as: bills of quantities, schedules of activities/works, schedules of rates, builder's quantities, variation accounts;

3 Valuation of variations using tendered rates, fair valuation/rates for new items of work, quotations, or dayworks;

4 Agreement/negotiation of variations. Documentation must clearly show the candidate's involvement with the piece of work and how they dealt with matters such as:

- design queries/Q&A sheets;
- to take lists;
- quantity checks;
- building up rates from first principles;
- inflation;
- prime cost and provisional sums;
- preliminaries, overheads and profit within variations; and
- professional and other fees within variations.

The following example commentary for the above competency relates to a bill of quantities submitted as evidence by the candidate.

How the competency is demonstrated

This bill of quantities demonstrates my knowledge and understanding of the principles of the quantification and costing of construction works and demonstrates how I can apply this knowledge by using the standard method of measurement and cost analysis. It shows that I can carry out measurement and costing at the initial stage of a project.

Wider skills demonstrated

This evidence also demonstrates my knowledge and understanding of client care in terms of understanding the requirements of my client and then collecting data that meet these needs. In addition, it shows my skills in collecting, collating and storing data. It was necessary for me to work with others in preparing this bill of quantities and this required me to use good communication and teamworking skills. The document also shows that I work in a professional manner giving sound, accurate advice to my client.

Background

This bill of quantities was prepared for one of our retail clients in relation to a small retail unit that was to be let under a [type to be given] contract. I prepared the bill at the request of my Senior quantity surveyor with whom I worked closely. In order to compile the bill I obtained copies of the drawings and specification from the architect and used the Standard Method of Measurement [specific method to be stated], undertook a manual take off and built up the rates from first principles. I compiled the data and presented this to my Senior quantity surveyor and then put together a draft bill, which we then agreed together. I presented this to our client and discussed the contents with them.

I have been involved with putting together many bills of quantities for a range of different small and medium sized contracts mainly relating to retail development projects.

The example above is 300 words so you will see that the information you can provide will be relatively limited.

> **TOP TIP**
> Your evidence and commentary, as well as showing your technical competency, are a great opportunity to show your written communication skills. RICS recommend that you draft your commentary in a word processing programme and run a spell check before cutting and pasting into the MLE.

The important things to remember are:

- Link the evidence to the technical competency using the requirements section of the competency description.

- Link the evidence to the relevant mandatory competencies, identifying the knowledge and understanding that the evidence demonstrates. Use the requirements of the relevant competencies.

- Set the evidence in context. In other words, you should clearly set out who was involved, who the work was undertaken for and how much experience you have in this type of activity.

Summary

- All competencies should be demonstrated by your evidence.

- Your evidence must be real outputs from your on-the-job work or, for some competencies, can be work from an academic course; it must demonstrate your competence.

- Evidence can only be linked to one technical competency but you can upload the same evidence again if you wish to use this for another technical competency.

- A 300-word commentary must be written with each item of evidence for each competency. Your commentary should explain how the evidence shows your skills in relation to one technical and any relevant mandatory competencies.

- Your commentary should set the evidence in context.

5 Gaining work experience

The next two chapters will consider the period of work experience that you undertake prior to final Associate assessment. This chapter will mainly be relevant to 'enrolment ready' candidates (see page 10).

Gaining work experience

Assuming that you are not a direct entry or assessment ready candidate you will need to undertake additional work experience before being ready for Associate assessment. This will take place over the required period (see figure 2 on page 9). You will need to gain a range of work experience that will allow you to achieve the required and defined levels of competence for your specific Associate pathway. In order to be sure that you are achieving these levels of competence I would urge you to work with an Associate supporter (see page 11). Although it is not a requirement of Associate assessment I would recommend that your supporter not only ensures you gain the right experience but also makes regular assessment of your progress against the competencies. In order to help your supporter make this assessment and to keep you on track, I would recommend that you keep a record of experience gained in each competency area.

The Associate supporter's role

The Associate assessment should be considered to be a process of continuous training and assessment. Likewise, your supporter's involvement with your progress should be continuous, and your assessment and training plan must be continuous and progressive also.

Your Associate supporter will usually be from within your employer's firm and will ideally be your line manager. In some cases, you may choose to use a supporter from outside your organisation, selecting someone you feel can act as a mentor and provide good support within your training area. If you do use an external supporter you should advise your employer of this and seek their approval of the arrangements. It will be important that your employer is happy with this arrangement since if your supporter feels you are not gaining the necessary experience you need to be able to feed this back to your employer without any awkwardness. In this situation a process will need to be put in place to allow your supporter to understand the work are involved with and to assess your competence in undertaking that work.

It is likely that where your line manager is your supporter they will already be doing what is required to assess you, as this requires similar processes to those used for appraising staff. Assessment involves being aware of how you are performing in day-to-day activities and reviewing work you have produced. By observing you at work they will be able to assess some of the requirements of the mandatory competencies, such as working in a team, problem-solving and working to deadlines. By looking at work you have produced, they will be able to learn more about your technical and professional knowledge, as well as your understanding. From this, your supporter can begin to form a judgment of how well you are doing.

If your supporter is in another department, or you have joined them on secondment, they will have to work a little harder. They will need to arrange to see examples of your work and perhaps to discuss this with some of their colleagues.

Your supporter should carry out a dual role, one to advise and support you throughout the process, and the other to regularly assess your competence. They are not expected to train you as such but rather to make sure that the training you receive during the course of carrying out your work in the firm matches and is applied to the requirements of the Associate pathway you are following. They should organise and direct the training to ensure that this occurs, to avoid gaps in your experience. They should involve you in all relevant work-based activities and encourage colleagues to do the same. Your supporter should help you to choose relevant evidence to submit for Associate assessment too.

Your supporter does not need to be RICS qualified but remember that you will need a proposer when you submit your evidence for assessment who must be an RICS member. Preferably your supporter would also be an RICS member and be from the same pathway that you are following to membership. Your supporter and proposer can, of course, be the same person.

In brief, your supporter should:

- ensure, day to day, and week to week, that you are gaining experience and training in line with the competency requirements of your pathway;
- deliver coaching and training, or ensure that you receive training with someone else, for example, with another department in your firm;
- assess you, ideally at three-monthly and six-monthly intervals throughout your period of experience;
- ensure that the evidence chosen for assessment is relevant, appropriate and accurately reflects your involvement;

- assist you in the preparation of the commentaries supporting your evidence;
- provide support and encouragement, and generally be a good friend to you, including on matters of general health and welfare. Consider your supporter as your mentor – someone you can bounce ideas off as well as be assessed by.

The most important thing to recognise is that their involvement with your training and assessment should be continuous, across the full training period. They should not simply be 'dipping in and out' of the process.

The brief notes below give some ideas of the duties I recommend be performed at various stages of the process to the Associate assessment. You should treat this as a handy checklist – nothing more. A more detailed breakdown of the stages is provided the following pages.

Bear in mind that this is 'an ideal world' scenario. Business or other demands may prevent you from gaining the necessary experience within the period specified in your assessment plan. If that is the case then the training period will simply be extended and the regular assessments should be maintained.

> **TOP TIP**
> Although not compulsory, it is beneficial to have an Associate supporter; this person will be able to advise, guide and assess you throughout the course of becoming an RICS Associate.

Registration

Your supporter's involvement will ideally start before registration. Before or shortly after you register I would recommend that a structured training plan be put in place providing a mechanism for you to choose the most appropriate pathway and competencies and to provide a plan that will help you to reach Associate assessment in the required timescales.

I would suggest that this stage of the process would include:

- meeting with your supporter to discuss their role;
- choosing the appropriate pathway and optional competencies;
- developing a structured training agreement by planning your training and experience to ensure the competencies and levels of the chosen pathway will be met with a timetable for each stage of this process;
- putting dates in your diary for three-monthly reviews and making sure these are agreed between all parties required to attend;
- agreeing a target Associate assessment date and building this into the timetable.

Three-monthly reviews

Every three months your supporter should carry out an assessment of your competence. You should present a summary of your experience against each competency in advance of this assessment and your supporter should assess your performance against each competency and advise you as to evidence that may be relevant for assessment. A progress report can be used to provide feedback to you. A review of the next stage of training should also be made.

In brief, your supporter should:

- check that your training is progressing as planned: review the structured training agreement and progress against the various competencies of your chosen pathway;

- review your documentary evidence to select items that could be used for assessment;
- review your structured development activities (see page 104) to ensure you are making good progress in this area;
- complete a progress report giving you feedback on your performance to date; and
- review the structured training agreement and provide an action plan for the next three months.

On completing the minimum experience required

This is the earliest point at which you can submit your evidence for Associate assessment. You and your supporter should consider carefully whether your submission should be made at this point.

Your supporter should:

- confirm any outstanding competency achievements;
- review your structured development to ensure that a satisfactory number of hours have been achieved over a range of activities;
- review the evidence held in the Managed Learning Environment to check these are the best and most relevant examples of your competencies;
- check your commentaries for accuracy and professionalism; and
- help you prepare for the ethics module.

At this point you will need your Associate proposer to verify that you are an appropriate person for RICS membership.

Progressive assessment

One of the basic philosophies of the Associate assessment is the idea that candidates will learn skills and acquire experience progressively. Over time you will progress from having 'knowledge and understanding' of a particular area, on to gaining practical experience of that area under normal circumstances, and finally, in some areas, to possessing the ability to apply your knowledge in more complex circumstances and to evaluate a situation in a wider context.

Registration

Registration is the stage at which you tell RICS about your qualifications and experience and RICS tells you whether you need to gain more experience before Associate assessment. You will also need to pay your Associate assessment fee and join RICS as an Associate candidate.

At this point you will need to consider the competencies for your Associate pathway. These are available on the RICS Associate website (www.ricsassociate.org). You should take some time (ideally with your Associate supporter) to go through the relevant competency descriptions and decide how much experience you have in total that is related to some or all of the competencies and whether you already have evidence available from previous experience to demonstrate your competence in them.

You should consider whether you already have or, if not, have access to the breadth and length of experience to achieve all the competencies? Remember you do not have to assemble all your evidence in one go: you can build up your evidence over time. Your evidence will be assessed only when you have put the full portfolio together and are ready for Associate assessment.

You will also need to choose your pathway and your optional competencies and again these should be discussed and agreed with your Associate supporter.

Gaining work experience

After you have registered you will receive an assessment plan setting out the number of years experience that you will need to achieve. You may already have achieved some of this and if so this can be taken off the number of years given. You should then work with your supporter to put together a structured training agreement setting out in detail how you will gain the necessary experience and/or qualifications.

Structured work experience is a structured approach to the delivery of training and experience over any given period. All firms who have provided traning for graduate surveyors (APC candidates) should have a structured training agreement in place and this may also be appropriate for Associate candidates.

A structured experience programme should be written to formalise the intention of the parties to deliver (on the part of the employer) and receive (in respect of the candidate) the training requirements of the chosen pathway over an agreed period and to specified standards of competence.

The important component of a structured experience programme is that it can demonstrate the ways by which the requirements of the Associate assessment will be met. Programmes can be developed and approved at a national level (for firms with offices throughout the UK) or at a regional level (just for one or a few offices in one region). A programme can be developed for organisations rather than individual candidates and once approved can be tailored and used for all candidates. The programme must be used in conjunction with the formal RICS Associate guides; it does not replace them.

The following information should generally be included within a structured experience programme:

- the role and responsibilities of the Associate supporter;

- the responsibilities of the Associate candidate;
- information on the organisation and areas of activity;
- a statement of the organisation's commitment to training;
- information on how the structured development requirement (see page 107) will be met and supported;
- arrangements and timescales for monitoring the training programme;
- a commitment by the candidate to follow the Associate guidance, to keep all the required documentation up to date and to make the necessary arrangements to meet with their supporter for three-monthly reviews;
- a competency achievement planner setting out the chosen competencies and the timescales within which the supporter expects the candidate to achieve each competency (see figure 3);
- a competency monitoring table setting out the experience that will be available for each competency that the candidate needs to achieve (see figure 4).

Gaining work experience

[Insert candidate's pathway]	Yr 1	Yr 2	Yr 3	Yr 4
[Identify in what year the competency should be achieved]				
Mandatory competencies				
Client care				
Communication and negotiation				
Conduct rules, ethics and professional practice				
Conflict avoidance, management and dispute resolution procedures				
Data management				
Health and safety				
Sustainability				
Teamworking				
Technical competencies				
[Enter your technical competencies in rows below]				

Figure 3 Competency achievement planner

Competency description
Conduct rules, ethics and professional practice
Predicted timings
Year 2 achievement
How is the competency to be achieved
The candidate will be provided with all relevant documentation, study materials and access to the RICS website to research and appreciate the implications of the RICS Rules of Conduct, ethics and professional practice. They will study the general principles of law and the legal system relating to their chosen qualification pathway and country of practice. The candidate will have the opportunity to attend seminars and discuss the implications of the Rules of Conduct, together with the application of the law in their area of surveying practice. They will identify situations in their normal working day where adherence to the Rules of Conduct and ethics are required. The candidate will be mentored in the detail and implications of the RICS Rules of Conduct and ethics together with the law in their area of practice. They will be tested/encouraged in relation to the company's working practices and asked to give reasoned advice to their Associate supporter on matters relating to professional practice. Also, the candidate will be given the opportunity to attend a further detailed seminar on this subject.
Action
Supporter to plan and monitor with candidate. Arrange to purchase RICS DVD.

Figure 4 Competency monitoring table

Practical guidance

Your Associate supporter may like to speak to an RICS regional training adviser (RTA) to seek guidance on the development and management of a structured experience programme.

Employers who have recently trained Associate candidates will already have a training agreement in place and will be familiar with RICS requirements. A degree of tailoring will be needed for each individual candidate and a proposed timescale and selection of competencies. RTAs can advise on both how to set up a new structured training plan and how to tailor an existing plan for a new candidate.

A structured experience programme is a demonstration of your employer's commitment to Associate training and may be used when advertising for new staff. It is, if you like, a kite-mark of excellence. Additionally, of course, having a plan in place saves time when considering the training plan for each new candidate.

It can seem as if larger employers will stand a better chance of getting candidates through the Associate assessment as they may have more opportunities for the candidate to gain a breadth of experience. In fact, everything depends on the firm itself. If an employer understands what needs to be done, and takes time to tailor requirements to the candidate's needs, then size is no issue. A smaller employer, where the candidate is in closer proximity to all other members of staff, and to the decisions being made, may have more scope for flexible development than a candidate in a larger, more impersonal, firm where the training programme may be more rigid.

Whatever the size of the practice, Associate supporters should take a close interest in your progress. Whether this happens in a large or small firm, this approach is the best possible training ground.

Remember that registration is the start of your journey to RICS membership. You should the meet with your supporter to discuss and agree the competencies to be achieved, considering your chosen pathway to membership and the resources of your firm. This is also the point to discuss how the training will be complemented by structured development (see chapter 6).

Overall, it is important that you discuss and agree expectations – your own, and your supporter's. Make sure that all parties start with a clear vision and understanding of the events that will follow.

End the process by putting dates in your diaries for the necessary reviews, and make sure that these are agreed by all the parties required to attend. This ensures a proactive approach to managing your training, creates a good impression, and should ensure that the dates do not slip.

Three-monthly reviews

The objective of the first three-monthly review is to allow you to gain the confidence that you have got off to a good start with the training period. A lot of your supporter's views and opinions concerning your progress will come from their observations, discussions and contact with you during the early part of the training period, which is itself a form of continuous assessment.

Care is needed at this early stage. All candidates develop at different rates, and their speed of development may vary from competency to competency. Sometimes candidates will make a slow start until they grasp or become familiar with basic concepts, but may then 'take off', with a rapid rate of development. The reverse may also apply, with candidates quick to pick up on initial points, but slower to transfer these into practice. Other candidates will follow a more consistent or 'straight line' rate of development. Don't worry about this – we are all different. Some candidates will also take a while to build confidence in their own ability and will appear slow initially, perhaps checking and double checking their work and what is required of them. It is important to develop your confidence and ability in your own time. Try not to compare your development with another candidate.

You should set aside a time and a place where you will be undisturbed for an hour or so. These reviews are very personal, privacy will be important.

The preparation of a summary of your experience in advance of the review will help to focus your mind, and will allow you to discuss how you feel your training is going and what progress has been made. This should ensure that you give your supporter a very honest opinion and view.

Your discussion in the review will need to relate specifically to progress against the competencies. You should get into the habit of bringing some examples of your work to either illustrate or remind your supporter of the standard of work you have achieved and to consider together whether these are appropriate documents for assessment. Your supporter will try to 'stretch' your knowledge and experience and may ask what you have learned from a particular situation, what problems were faced, and how they were resolved, and how the experience could be transferred or adapted to deal with other areas of work. This ability to learn, and then transfer skills and experience to address other problems, is an important facet of being a professional. It also allows your supporter to check that you have a good all-round view of any particular situation or technique, enabling you to repeat your actions under different circumstances.

At the end of each three-monthly review, your supporter should firstly confirm any competencies that they feel have been achieved and agree with you the items of evidence you will use to demonstrate this. They should also complete a progress report giving some notes concerning training to date, experience gained and your abilities. At this first stage, your supporter's views concerning your abilities may be very general; however, you should keep an accurate record of your discussions around the competencies – later reviews will be able to draw upon more information. You should also add some brief comments to the progress report.

After filling in the progress report, it should be signed and dated by you and your supporter as a formal record of your assessment. These progress reports are not submitted to RICS but are a very useful record of your training and development.

An example of a three-monthly progress report is shown on page 101.

It is always useful to record a few action points in the progress report as well but be realistic about what has been achieved in the first three months – there is plenty of time yet!

At this stage you should also discuss your structured development to ensure you are making good progress in this area. You would expect to see around 12 hours recorded at this stage.

Note that the process carried out in the first three-monthly review should be repeated for subsequent reviews.

Your supporter should establish a rigorous approach to the assessment of the competencies from the outset and retain this throughout the course of your experience. Many candidates have expressed concern that supporters do not assess competence rigorously enough during the structured training. This inevitably leads to a lack of confidence as the candidate is not sure whether they have truly met the levels of competence that are to be confirmed by their proposer prior to the final Associate assessment. This can sometimes be taken with them right through the Associate assessment and can have a seriously detrimental effect on performance. It is crucial that competency achievements are 'hard won' although, of course, not unachievable!

Sample supporter's progress report (residential survey and valuation pathway)

To be completed every three months.

Date: 30 March 2011 (Period January 2011 to March 2011) – Observations on training to date, experience gained and ability of candidate:

Training to date

Julie has attended a two-day teamworking course. This course has added to her ability to get on well with a variety of work colleagues and to appreciate their styles of approach.

Julie has built up a solid base in her first three months. In particular she has been able to produce financial appraisals and liaise with sub-consultants on such matters with the minimum of supervision.

Report writing – soft market testing report, cash flow report

Cash flow calculations – cash flow appraisal

Brochure wording and questionnaire

Evaluation process

Market testing exercise with lead developers

Involvement on report editing

Research – Julie has researched market evidence for residential space in connection with regeneration projects

Client contact – client exposure has increased dramatically and Julie has handled clients well.

Ability of candidate

Interpersonal skills – Julie is a good team member and fits in well. She has a good rapport with clients.

Written material – Julie has an appropriate use of business English and presents information in an organised manner.

Numerical – Julie has an excellent understanding of figures and the ability to use relevant systems and create cashflow spreadsheets.

Self-management – Julie makes good use of the working day to earn fees and has absorbed quickly the fundamentals of drafting and issuing invoices.

Competencies achieved at date of review:

Communciation and negotiation

Teamworking

Signature: N Jones Date: 30 March 2011

Candidate's comments:

During my first three months within the Public Sector Consultancy team I have worked on a number of interesting large-scale local authority development schemes. The majority of the work carried out has been in the form of development appraisal analysis and other numerical tasks. I am thoroughly enjoying my work within the team and am continually learning from the other members of the team.

I feel a valued member of the department and believe that I have shown many of my strengths in the work that I have undertaken. During the next three months I hope to gain more experience in some of the areas in which I am not so competent, such as letter writing and report drafting. Also, during this period I would like to gain experience in general areas of surveying, such as landlord and tenant issues.

Signature: Julie James Date: 30 March 2011

Actions at the halfway point

At the halfway point of the period of structured training it is important to have a strategic review of progress and to plan the remaining months of training. This should be undertaken alongside one of the three monthly reviews.

You should identify and discuss any gaps in experience. Any such gaps will form the basis of your training and experience needs over the remaining period.

Your supporter should then review the structured training agreement, in particular the competency achievement planner and monitoring table, to reflect any changes needed to the plans for your training.

You should also review your structured development with your supporter and ensure you are on target to achieve the required number of hours.

Following completion of your minimum period of work experience

At this point you should review with your supporter whether you have achieved the necessary competencies and have sufficient and appropriate evidence held in the Managed Learning Environment to support this. Consider also whether you yourself feel ready to become an RICS Associate.

Summary

- It is important that your work experience is structured around the competencies for your pathway.
- You should work closely with your Associate supporter both in setting up your training agreement and in reviewing your progress.
- Encourage your employer to set up a structured training plan using a competency planner and a monitoring table.
- Organise three-monthly progress reviews with your Associate supporter.

6 Structured development

Structured development is a type of professional development. Before we consider the requirements for Associate assessment in detail let's take a look at professional development generally.

The concept of professional development

Professional and personal development is an essential part of being a professional. Creation of a competent, innovative, forward-looking profession and professionals is achieved by proactive and targeted learning.

Throughout your professional career you will need to undertake and record appropriate professional development (or lifelong learning as it becomes known when you qualify) and may need to provide evidence of having done so to RICS. Structured development to support the attainment of the Associate competencies is the first stage of this.

The aim of professional development is to:

- improve your professional competence; and
- demonstrate how you have maintained professional competence.

You can demonstrate that you have complied with this if you plan, carry out and record an annual program of learning activities designed to maintain competence and improve professional knowledge.

The idea behind professional development is that it provides the opportunity for you to acquire some of the additional skills and knowledge that it will not always be possible for your employer to provide within the week-to-week business of the practice.

An important aspect of your professional development is that it should be planned and structured in such a way as to remain flexible. It should be designed to complement and support your training and development in the context of the various competencies.

Professional development may comprise formal training courses or more informal types of learning, such as structured reading, distance-learning or e-learning programmes, and secondments. It is important that you accept ownership of your professional development, recognising that the planning, acquiring and evaluating of it is your responsibility. Professional development should:

- be gained in a structured manner;
- be based on an explicit process of selecting, planning and evaluating the activities; and
- reflect learning from informal training sources such as structured reading and secondments.

Remember that your learning and development never ends. Your professional competence will continue to be assessed by employers, clients and peer groups throughout your career. As the world around you changes in terms of consumer demands, law and technology, the content and focus of your learning and development will also change.

Therefore, when you become qualified, think of the RICS programme of continuous professional development

(CPD) and lifelong learning as a vehicle to assist you with your essential and ongoing training and development needs, in a rapidly changing environment.

Structured development for Associate assessment

For Associate assessment you are required to keep a log of structured development activity in the Managed Learning Environment. Your structured development record is a record and evaluation of the learning activities that have built up your skills towards your mandatory and technical competencies. Structured development is private learning, organised learning, work-based learning or other activities that you undertake in order to reach the required standard for your qualification. It should be gained in a systematic, structured manner and should be based on a process of selecting, planning and evaluating the activities.

I would recommend that you create a structured development plan focusing on the following four-stage cycle:

1. Appraisal – assess your current situation and identify your learning needs.
2. Planning – set your learning objectives.
3. Development – find and particpate in suitable activities.
4. Reflection – reflect on your progress.

You should start by asking yourself the following questions:

- Where am I going?
- How will I get there?
- What will I need for the journey?

- How will I know when I have arrived?

Your answers to these questions should produce the six critical elements of an effective structured development plan:

- A clear statement of where you want to be at the end of a specified period.
- The specific knowledge and skills you will need to get there.
- The actions you will have to take to acquire the knowledge and skills you will need.
- The resources and support you will require to attain your goal.
- The criteria by which you will determine whether or not you are still 'on course'.
- The intermediate stages with dates for completion and/or review.

You may like to download the RICS Personal Development Planner from the RICS website (www.rics.org/cpd) to help you with this.

Think in broad terms about your experience and qualities. Do not limit yourself to technical skills alone; take time to consider the mandatory competencies and other wider skills you will develop.

Stage 1 – Appraisal

Where am I now?

In order to identify future learning and development needs it is important to review your personal and professional experience to date. It is often said that there is no point in deciding where you are going until you have established where you are now.

Structured development

As with other areas in business, identifying what you have already achieved (in this case in terms of skills and knowledge) can provide a sound basis for planning for the future.

An effective method of self-appraisal is to identify your strengths (S) and weaknesses (W) and to examine both the opportunities (O) and the threats (T) you may face.

'SWOT' Analysis

Strengths	Weaknesses

• What are your core skills?
• What do you do well?
    – technical skills and knowledge?
    – other transferable skills?
	• Where are your skills/knowledge lacking?
	• What would you like to improve?
	    – from your own point of view?
	    – from the point of view of other people?

Opportunities	**Threats**

• What are the opportunities facing you?
• What are the interesting new trends?
    – changes in markets and professional practice
    – emerging new specialisms
    – developments in technology
    – moves towards quality assurance
    – assuming a management role
	• What obstacles do you face?
	• Is your role changing?
	    – competition from other businesses
	    – merger with other bodies
	    – legislative changes
	    – different skills required when running a small business
	    – limited opportunities for progression
	    – threat of redundancy

This kind of analysis should enable you to determine areas of interest and ambitions that can be used to shape plans for further development.

Although there are clear benefits in planning structured development to develop knowledge and skills in new or weaker areas, you should not overlook the potential for further development in your stronger areas. Building on existing strengths is as relevant an aim for structured development as improving in areas of weakness.

Stage 2 – Planning

Where am I going?

Having established areas for action, the next step is to detail your priorities for development.

Following the completion of stage 1, you should be able to identify gaps in your skills and knowledge where you can set specific development objectives. There should be a clearly defined relationship between your structured development and your Associate competencies. It is crucial that your objectives will lead you to achieving the Associate competencies.

These objectives should contain an element of challenge so that they carry you on to new ground, but they must also be realistic. At this stage it is useful to set targets in terms of required levels of competence.

Description of skills, competence and experience

What best describes your current level of competence and the level you now want to attain?

Unaware
- unaware of subject area and knowledge;
- possess little or no knowledge/skills;
- require full training and development.

Aware
- possess basic knowledge/skills;
- unable to work without supervision;
- require training/development and more in-depth information.

Capable
- possess adequate knowledge/skills;
- able to work with some autonomy;
- able to work effectively as part of team;
- require guidance/some further training.

Skilled
- possess requisite knowledge/skills to perform effectively and efficiently;
- able to work with considerable autonomy;
- need occasional top-up training.

Expert
- acknowledged by others as an authority;
- very substantial personal autonomy.

When establishing your objectives, you should also work within practical constraints that may influence methods of development. Factors you may wish to take into account include:

- What opportunities and support for learning are available?
- How much will it cost in terms of money, time, and conflict with other commitments?
- What added value will result – qualification, promotion, new business?

Finally, objectives should be set within a realistic timeframe. In some cases they will not be easily achievable within a 12-month cycle. However, it should be possible to determine some progress towards achieving an objective in this time period and to re-evaluate long-term objectives in the continuing cycle of development.

Stage 3 – Development

How will I get there?

The achievement of development objectives requires involvement in a wide range of learning activities on a continuous basis. Structured development may take the form of any appropriate learning activity and need not necessarily be biased towards course attendance.

The decision as to what constitutes relevant learning and development must lie primarily with you as an individual. Many different activities can qualify as structured development. Deciding on what is relevant is up to you because learning is based upon personal assessment.

Do not restrict your learning to formal training courses, seminars and workshops. RICS, like many other professional bodies, recognises there are many other relevant methods of learning.

When arranging your structured development activities you will need to consider your preferred learning style. Individuals perceive and process experience in different ways and these differences comprise a unique learning style. An awareness of your preferred ways of learning is therefore essential to the quality of learning undertaken.

The four major learning styles are highlighted in the chart below, which may help to guide you in the choice of activities.

Learning styles

Innovative learning:

- connect new information/skills with personal experience and real-life problems/situations;
- prefer co-operative methods of learning, for example, seminar groups, brainstorming, learning through project work.

Analytic learning:

- want to acquire knowledge to deepen understanding of concepts/processes;

- prefer to learn from what 'the experts' have to say, for example, lectures, conferences, further qualifications.

Common sense learning:

- be interested in how things work and want to 'get in and try it';
- prefer experiential methods of learning, for example, through hands-on tasks, on the job learning.

Dynamic learning:

- rely on self-directed discovery and want to teach yourself;
- prefer independent study and training which involves simulations and role-play.

Many people will find that they have a mixed learning style that involves two or more of the above. This situation provides a wide variety of possible methods for effective structured development.

Although the emphasis of CPD should be on planned activities, you should also be able to recognise and use unplanned opportunities that may arise.

If you are involved in an activity that did not feature in your development plan but from which you have learnt something, record it and consider whether it can apply to any of your specified objectives or to a new objective in the continuing development cycle.

Stage 4 – Reflection

How will I know when I have arrived?

To gain the full benefit from any structured development activities undertaken it is necessary to evaluate the outcomes and to establish whether you have achieved your objectives.

When reflecting on your activities you should consider whether you have experienced personal or business benefits from your efforts through the practical application of what you have learnt.

Evidence of such achievement can be demonstrated in various ways as illustrated below.

Evidence of skills acquisition and improved competence

Recognised qualifications
- short course completion certificates
- credits for accumulated qualification
- NVQ
- Diploma
- Certificate in Management Studies (CMS)
- Academic qualification

Self
- measured against own criteria
- discussion with manager/colleagues
- favourable annual appraisal
- recommendation for promotion
- change in professional role/duties

Organisation
- adoption of recommendations as policy
- improved business performance
- cost efficiency savings
- safer working environment
- achievement of quality standard

Colleagues (peers/superiors)
- request to coach/advise colleagues
- suggested to join/lead project team
- request to compile paper/manual

Public
- membership of other professional groups
- solved problem for community group
- publication of papers/research
- request for advice regarding policy/law

Client
- award of further work
- award of commission in new area
- recommendation to other organisation

The evaluation stage deserves special attention as it produces a summary of achievements that demonstrate how you have met your original objectives. For any areas of under-achievement you should reconsider whether the target originally set remains valid or whether you require more time to achieve it. By reviewing the outcomes of your structured development activity in this way you will continue the learning and development cycle into the next year and beyond.

Types of structured development activities

When uploading your structured development activities on to the Managed Learning Environment you will need to identify for each activity whether it is personal, work-based or organised.

Personal

This may include reading, online learning or similar activities that you have undertaken independently. Structured reading is one of the most efficient and cost effective ways for members to broaden their knowledge in areas of general professional interest and also in more specialised areas. There are savings to be made from using this method both in terms of time and money. Suggested sources of reading might include:

- books, journals, magazines, newspapers;
- internet sites of professional interest;
- technical manuals;
- research papers;
- CD-ROM disks, videos, audio tapes;
- transcripts of speeches/conference items, lecture/seminar notes;
- reports/guidance notes issued by professional bodies;
- CPD study packs.

The value of learning through reading about the experience and the advice of others is widely perceived to be one of the key tools in the acquisition of professional knowledge. This concept primarily concentrates on the need for a structured and disciplined approach to private study. Ad-hoc reading of trade/professional journals can be useful in updating basic knowledge, but only to a limited extent. Professional structured learning should go much further than this.

An in-depth study on a given subject area, using several different references if possible, should increase the level of knowledge and provide a balanced view of the issues involved. Such study could form a programme, particularly if based on an imaginative use of references. It is important to remember that the greatest benefit can only be achieved through using this material in a structured way. This could include the following actions:

- Read an article on an identified subject or new topic, if relevant.
- Search the RICS Library catalogue for professional group reading lists. Check whether you can borrow

Structured development

the material you need or purchase direct from RICS Books (www.ricsbooks.com).

- Do not overlook the information available through the Owlion audio cassettes or CPD study packs available from the RICS Library (www.rics.org/library) free of charge.
- Write up your notes and findings as a reference document for your company so that others may benefit from your research and consider whether it would be worthwhile to arrange a discussion with colleagues.
- Prepare a talk to give within your company or to a local group.
- Prepare an article for publication in Modus or your regional newsletter.
- Record the experience in the Managed Learning Environment.

Each stage can count towards your structured development.

Private learning could also include activities outside of your work place. Skills acquired from personal activities outside the workplace may be just as relevant to the development of personal competence and are a valid source of structured development where you feel they help you to achieve the Associate competencies. For example:

- voluntary (non-professional) work – active involvement with voluntary/charitable organisations can be a rewarding experience. You may be learning about applications for grants and lottery funding or developing skills in organising/running meetings;
- school governor, Justice of the Peace, member of Parent Teacher Association – these activities can

broaden knowledge and understanding of many issues that form part of the broad context of people management;
- parish council/other local government work;
- youth groups;
- housing association or other board or committee; and
- local business community boards, for example, Chamber of Commerce.

Organised learning

This may include a learning event provided by a training company, college or similar, or may be a CPD event run by RICS or another professional institution. Other activities can include long-term qualifications and projects (six months or more), for example full/part-time study, open distance learning, or contributions to original research.

Any organised learning recorded as structured development should be in addition to any study undertaken for an academic or vocational course that would exempt you from a period of work experience for Associate assessment. For example, if you are taking a HNC this should not be recorded as structured development.

Work-based learning

This would include training provided in your workplace. This may be in the form of in-house training courses or events put on by your employer; instruction or mentored practice in new tasks; or reading, study or online learning required by your employer in order to equip you for your role. Work-based learning can also include a range of other professional activities that are linked to your work. The following list provides some ideas of what you might do:

- on the job development:
- business management skills;
- coaching/mentoring;
- personnel management skills;
- planning and running an in-house training event;
- internal discussion groups;
- adviser/consultancy positions;
- arbitrator/expert witness/adjudicator;
- professional interviews;
- voluntary (professional) work;
- special project work;
- staff training;
- study of a foreign language (this may be relevant if you work in an international context);
- information technology; and
- secondment – transfer to another department;
- presentations and publications:
- research for publications and papers;
- preparation for presentations to colleagues, clients, local groups;
- participation in public meetings; and
- lecturing on careers/the profession.

There are, of course, many other activities that could be used towards your structured development requirement. Remember that you must have recorded a minimum of 48 hours structured development in the 12 months prior to your Associate assessment and be mindful that this is a minimum and the greater range and quantity of activities you can undertake the better.

If you are ready for Associate assessment now, you can complete your structured development record (on the MLE) by reviewing your learning activities over the last year (if you have no diary records of your learning activities you may have to use approximate dates). If you are working towards Associate assessment in the future, you should complete the record as you go. It will be important that you are able to reflect on the learning you have undertaken and the detailed process above will help you with this.

There is no strict rule about the precise number of hours of structured development you record for each individual competency. However, you should ensure that you achieve a reasonable spread of hours across the competencies, and record a variety of activities and learning methods. You do not have to record something for every competency and some of your activities can relate to more than one competency.

Completing the structured development record

You complete your structured development record by typing direct into the MLE. To do this you access the structured development recording area. Follow the instructions on screen to log your activities. Detailed guidance on the process is in the MLE guide available in the official Associate pathway guides and on the MLE.

You should not log any activity that took less than half an hour but you may be able to put a number of small activities together if they of a similar nature in order to build these up to a full half hour. You should start a new entry for each activity.

You must follow the prompts to record:

- a brief description or title – for example, 'event to learn about new forms of contract';
- the start date and time – when you commenced the learning activity;
- the end date and time – when you completed it;
- a description of the activity – for example, 'lecture at [venue] on the subject of … [description of what the lecture covered]'. Make sure the description relates directly to the competency: show how it was relevant and useful;
- an activity review – reflect on what you have learned and describe the learning outcome; for example, 'raised level of skill from basic awareness to a good working knowledge'. This would build from the last stage of the cycle discussed on page 113.

You will then need to save and link to the relevant competencies (see page 78). The MLE guide also shows you step by step how to do this (see page 129).

The following are examples of the information you may provide.

Example structured development record

> **Title:** Different types of construction contract
> **Start date/time:** 14 May 2011 14:00
> **End date:** 14 May 2011 16:00
> **Type:** Personal
> **Details:** I reviewed the pages on isurv to gain a full understanding of the different types of construction contract, how to select the most appropriate form of contract and to review case law, in particular those relating to disputes over contract terms. I then considered the different construction projects that my company are currently involved with and wrote a short essay on what form of contract I thought would have been most appropriate and why. I reviewed this with my Associate supporter.
> **Activity review – objective:** The objective of this activity was to develop my understanding of the many different types of construction contract, the circumstances when each of these is used and the disputes that can arise.
> **Activity review – outcome:** Before starting this activity my knowledge of different forms of contract was 'basic' as I had defined in my structured development plan. After consultation with my Associate supporter and from his feedback this has now increased to 'aware'. I now need to undertake further study to increase this to 'competent'.
> **Number of hours:** 2
> **Link to competencies:**
> *Technical:*
> Contract practice
> Procurement and tendering
> *Mandatory:*
> Conflict avoidance, management and dispute resolution procedures

Structured development

Note: If the discussion with your supervisor had been at a separate time you would need to record this as a separate structured development activity. The times you record will be used by the MLE to calculate the number of hours of activity undertaken so should relate to a continuous period of learning.

Title: Report writing
Start date/time: 18 June 2011 10:00
End date: 18 June 2011 14:00
Type: Organised
Details: I attended a report writing workshop run by Innovate Training for my company. This involved both presentations on best practice and practical sessions where I practised my report writing skills. I also spent time after the formal event discussing key issues with my colleagues.
Activity review – objective: The objective of this activity was to develop my understanding of report writing, the different techniques that can be used and how to make sure that my reports meet the requirements of our clients.
Activity review – outcome: Before starting this activity my knowledge of report writing was 'aware' as defined in my structured development plan. I now feel that following this workshop my knowledge and understanding is 'competent'. I will now make use of this training in my everyday work.
Number of hours: 4
Link to competencies:
Mandatory:
Communication and negotiation

Summary

- Structured development is a type of professional development that you will pursue throughout your professional career.
- Structured development should be planned and structured around your strengths and weaknesses and you should set objectives against which you can then reflect on your learning.
- A range of activities can be included as structured development. These should be categorised as either personal, work-based or organised learning.
- Structured development must be linked to the Associate competencies.

7 The Managed Learning Environment (MLE)

I have referred to the Managed Learning Environment (MLE) a number of times already in this book. In this chapter further advice is given as to the structure and use of the MLE as a mechanism for demonstrating your competencies. Before we look at the Associate MLE lets take a moment to stand back and consider what is meant by this generally.

What are managed learning environments?

Managed learning environments (MLEs) are software systems that are configured to help a facilitator manage an education and/or assessment process. This is in contrast to a virtual learning environment, where the focus is on the tools used in the process of teaching in an online setting. Examples of virtual learning environments include Blackboard (software application for education institutions) used by many universities and colleges offering surveying courses. Blackboard offers opportunities for online discussion and also to hold presentations and other resource material. Another frequently used programme is Moodle (Modular Object-Oriented Dynamic Learning Environment). Essentially, a managed learning environment helps to set the perimeters of the learning experience, and makes it possible to standardise the resources used in learning in a manner that results in measured results.

While originally envisioned as the framework for taking distance learning into web-based learning, the managed learning environment software has become increasingly versatile and suited for use in offline applications as well. At the same time, the software has become more sophisticated and user friendly, making it a viable tool for use in a number of different learning and assessment situations.

The RICS Associate Managed Learning Environment

The RICS Associate Managed Learning Environment (MLE) is an online resource that enables you to upload evidence, write your 300-word commentaries and record structured development for the Associate assessment. It is accessible from any computer connected to the internet.

Once you have entered the MLE you can progress at your own pace. The MLE will give you tips and guidance on the screens. In the MLE you will build up, piece by piece, evidence to show your knowledge and skills. When you believe you have met all the required competencies, you apply for Associate assessment.

RICS Associate assessors will view all your evidence and decide whether they agree that you have met the requirements to become an associate. If successful, you will be awarded the AssocRICS designation; if you are not successful, you will be given feedback about what you need to work on. You will then collect further evidence as directed and resubmit for Associate assessment.

The requirements fall into three:

- written evidence – examples of written work prepared by you, during the course of your day-to-day employment, submitted electronically to the MLE;

- commentary – with each piece of evidence you will write a 300-word commentary; this is typed directly into the MLE;
- structured development – an account of your learning activities and outcomes over the last twelve months, explaining what you have done in order to build up your competence; this is typed directly into the MLE.

The MLE has additional functions to help you keep track of your uploaded evidence and structured development:

- alerts – to notify you of any important changes and developments, new guidance notes, or new learning;
- details of additional resources available to help you with your Associate assessment;
- events – RICS will use this section to advertise events that could count towards your structured development.

The MLE in detail

In the following section we will take a look at the MLE in some detail and consider each of the different sections. You should, however, refer to the MLE User Guide for definitive advice on how to use the MLE. This section is simply intended as an overview.

Homepage

The homepage contains five panes: My Details; Welcome; Alerts; Events; Diary.

The My Details pane displays your name and, if necessary, the number of years' relevant experience you are required to complete before you can apply for Associate assessment. This pane is fixed in position and is always visible.

The homepage allows you to arrange the other four panes to suit your preferences; simply click and drag the title bar of the pane. The panes can also be collapsed and expanded as necessary by clicking on the triangle symbol on the right of each title bar.

The Welcome pane contains links that take you to the most frequently used areas: Upload my qualifications; Upload some new evidence; Record my structured development; Download the guidance notes. There is also a link to submit your evidence for assessment.

Alerts are the notifications from RICS of any important changes. Clicking on the title of the alert opens the full message.

The Events pane displays a selection of current RICS events that can help you develop your competencies. If you click on the event title, you will be taken to the full event details. To subscribe to an event, click through to the event website. To see a full list of RICS events, go to www.rics.org/events

Finally, the Diary allows you to view on what dates you added your structured development, evidence and any event reminders you have added.

Figure 5 shows the homepage.

Selecting your competencies

The very first time you log into the MLE you will be asked to select your optional competencies. If your pathway has a choice of competencies, this is where you will choose the ones you want to submit evidence against.

You can change your optional competencies at a later date if required. To do so, click 'My details' in the header bar and then select 'Optional competencies'. If you change options, you must reassign or delete any evidence you have logged against the original choice before you can go forward for assessment. You must have the correct number of pieces of evidence assigned to competencies before coming forward for assessment; any more or any less and the system will not allow you to progress.

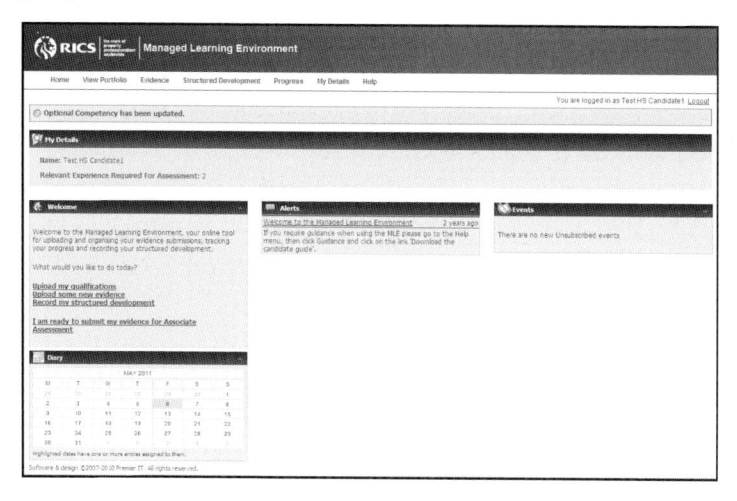

Figure 5 MLE homepage

Also when you first log in you must complete your work experience record. From the homepage click 'My details' and then 'Add work experience'. Your work experience must be complete and up to date when you apply for Associate assessment.

Adding evidence

The Add evidence screen in the MLE is where you upload your evidence and add commentaries. For each piece of evidence you will need to input:

1   Title – this is the title of your piece of evidence for submission. You should try to make it as clear and descriptive as possible (character limit of 100 applies).

2   Evidence summary – this is a free text area where you provide your commentary on the piece of

evidence. There is a character limit of 300 words for each piece of evidence. You should not use any abbreviations or shorthand in your commentary. Remember the commentary is an important part of your evidence (see page 78).

Figure 6 below shows the Add evidence screen.

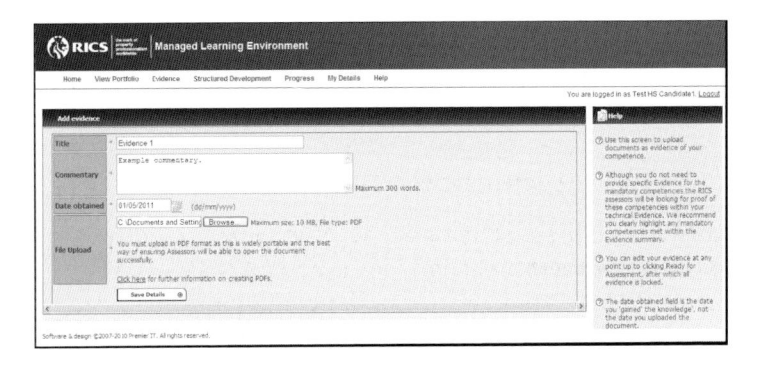

Figure 6 Add evidence screen

Once you have saved your evidence, you will be taken to the screen that allows you to assign that piece of evidence to your competency. The first screen shows all the technical competencies you are required to achieve for your chosen pathway. You will need to click in the box next to the title of the competency, and the piece of evidence will be linked to that competency. You can link each piece of evidence to only one competency (if you also consider it relevant to one of the mandatory competencies this should be explained in the commentary).

Confirmation of evidence submission

You will be able to view a summary of the evidence you have assembled on the Summary of evidence screen (see figure 7 below).

The Managed Learning Environment (MLE)

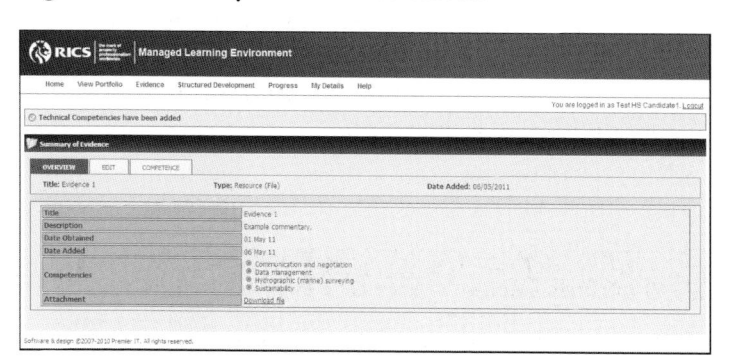

Figure 7 Summary of evidence screen

Structured development screens

You can access the structured development recording area through the homepage via the link 'Record my structured development' in the Welcome pane. You can also access it by clicking 'Structured development' in the header bar and then selecting 'Add structured development'. For guidance on what should be included in your structured development record see page 115.

You will need to enter the following details for each of your structured development activities:

- Activity – enter here a brief description of the structured development activity (for example, 'attended an event about...'; 'studied an online module on...').
- Start date – enter the date and time you started the learning activity.
- End date – enter the date and time you completed the learning activity.
- Type – select the category of learning activity (see page 115 for further advice). The options are:
    - personal learning;
    - organised learning;

- work-based learning; and
- other.

- Description – this is an overview of what the learning entailed. You must include your learning need and details of the learning activity. See chapter 6 for advice on planning your structured development, from which you will be able to identify your learning need.

- Add activity review – this tick box will only appear if the start and end date are in the past. You can put future events in as structured development but the system will let you know, when the date is past, that you need to add an activity review. The activity review is where you record the time you spent on the activity and reflect on what you have learned. You must describe the learning outcome and relate it to the competencies.

When you click 'Save', you will be able to assign your structured learning activities to the competencies.

Figure 8 shows the 'Add structured development' screen.

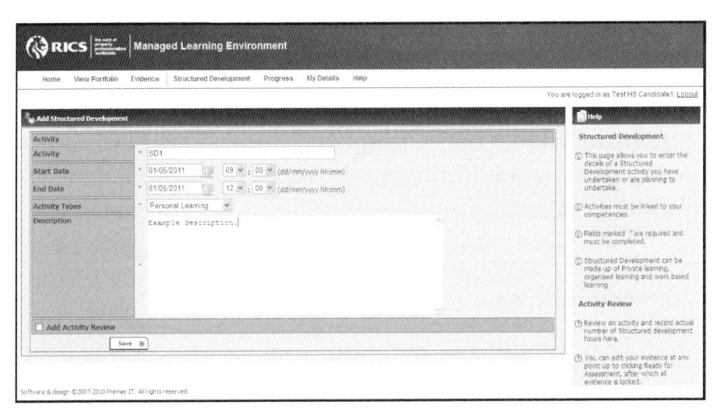

Figure 8 Add structured development screen

A confirmation banner will appear below the header bar to let you know that your structured development record has been saved.

The Managed Learning Environment (MLE)

The next screen requires you to assign the structured development activity to any relevant technical competencies; you can assign it to any number of the technical competencies or you may choose not to assign it to any. When you have finished, click 'Save'.

The following screen requires you to assign the structured development activity to any relevant mandatory competencies; you can assign it to any number of the mandatory competencies or you may choose not to assign it to any. However, if you did not assign the activity to a technical competency you will need to assign it to at least one mandatory competency – each structured development activity must have at least one competency association.

You will be able to edit entries or assign or remove competencies from the list of structured development activities screen. To amend any entry, you simply need to click on its title. See Figure 9 below.

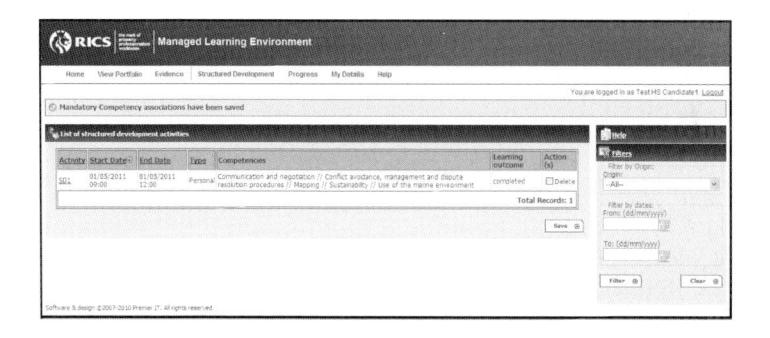

Figure 9 List of structured development activities screen

After you have actually completed the activity you will need to add an activity review including the outcome of the learning and how many hours of structured learning it represented. See Figure 10.

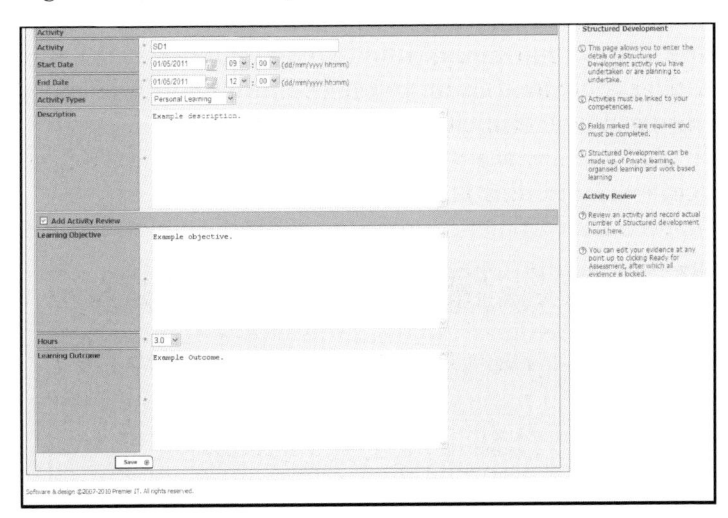

Figure 10 Add activity review screen

You will be able to view a list of the structured development that is ready for assessment, on the 'Structured development summary'. You will only see the structured development for which you have already added an activity review, as only these entries are ready for assessment. You will see a total number of hours completed at the bottom of the page. To amend any entry, you simply need to click on its title.

Viewing your portfolio

You can view your portfolio by clicking 'View portfolio' in the header bar. There are four dropdown menus:

- Technical competencies;
- Mandatory competencies;
- Type;
- Diary.

Under technical competencies, if you click on a competency it will expand, showing you what evidence has been submitted against this competency. In green you will see any structured development you have submitted against that competency; in red you will see any evidence you have submitted against that competency. If you click on an entry you will either be able to view the file you uploaded or view the structured development. (See figure 11 below.) The same is true under mandatory competencies.

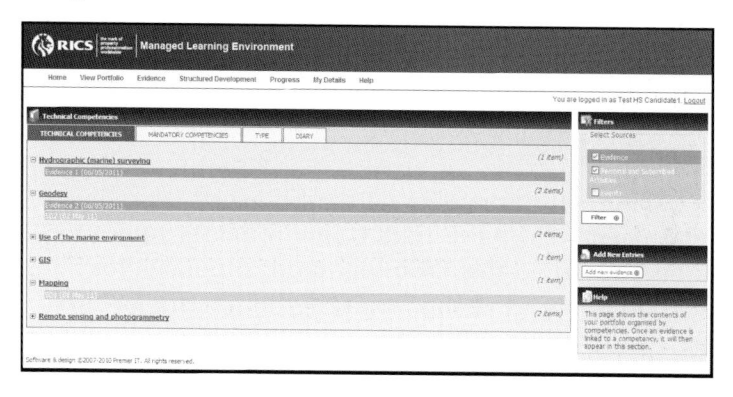

Figure 11 Technical competencies

Under the diary tab you will be able to view a log of what you have added by date. You are also able to view your overall progress by clicking on 'Progress' in the header bar. If you select 'List assessment' you can review your progress and any outstanding actions.

By clicking the 'Review progress' tab you can see how many hours of structured development you have added and how many more are required.

- Green – you have submitted all of the required pieces of evidence;
- Amber – more evidence is required;
- Red – you have not submitted any pieces of evidence against that competency yet.

All the competencies will have to display as green before you can proceed to Associate Assessment.

Ready for assessment

You will be ready for Associate assessment when you have completed the following:

- submitted the required number of pieces of evidence with commentaries (four for each technical competency);
- recorded 48 hours of structured development over the last 12 months;
- recorded your work experience; and
- uploaded your signed declarations.

You submit your portfolio via the Managed Learning Environment. The MLE User Guide provides advice for you on how to do this. You will need to upload your signed declarations at this point, if you haven't already done so.

The formal RICS Associate guidance explains in detail what declarations are required. If, when you registered for RICS Associate, you selected an academic or vocational qualification, you must upload it before assessment; you must upload the certificate or verification letter.

If you have added too many pieces of evidence for a competency it is overloaded. In this case you will need to go into the system and remove or reassign the evidence. You can only submit the required number of pieces of evidence (for example, 24 for the quantity surveying and construction pathway) and 48 hours of structured development. Any more or any less than this and the system will not allow you to progress. Once you have submitted your assessment you will not be able to edit anything within the MLE.

When making your submission for Associate assessment you should check all your details via the 'My details'

The Managed Learning Environment (MLE)

screen and complete your job position and workplace: this will ensure once you are ready for assessment there is not a conflict of interest with the Associate assessors.

The second two sections on the 'My details' screen allow you to view the pathway you registered for. You cannot change your pathway. If you do wish to transfer to another pathway, you should contact RICS.

You can also enter the details about your qualification and upload your relevant certificate. The title of your qualification is automatically populated from your original application through the registration pages. If you need to amend this, contact RICS. RICS will need to re-evaluate your registration if you provided inaccurate information at the registration stage. If the screen reads 'Not set' you did not enter any qualification when you registered for the process. If this is due to error, and you do have a relevant qualification, contact RICS.

You must select your practice area. This will enable RICS to match you with an appropriate assessor.

Once everything has been uploaded and all your details are complete, go back to 'Submit assessment for review'. The MLE will display a warning message telling you that when you have clicked on 'Continue' your evidence will be locked and you will be unable to add or amend anything that you have submitted as evidence. By locking the record, RICS ensures that the assessors are reviewing the correct version of documents and that nothing has been changed.

Your submissions will now be assessed by the Associate assessors (see chapter 9).

Summary

- Managed learning environments (MLEs) are software systems designed to manage an education and/or assessment process.

- MLEs are not intended as online classrooms.

- The RICS Associate MLE allows you to upload evidence, add commentaries and record structured development.

- The MLE also allows you to track your evidence and to see where you need additional documentation.

- The MLE allows your Assessors to access your evidence and to make a judgment as to whether you are passed or referred.

- In the unfortunate event that you are referred, the MLE allows the assessors to give you feedback to help you to prepare for reassessment.

8 Professional ethics

In this chapter I would like to give you some of the background information that will help you to prepare for the ethics module. We consider what is meant by professional ethics and what the RICS Rules of Conduct require. We also look at some issues for you to consider prior to taking this module.

Being a professional

As a professional body RICS has a responsibility to protect the public by ensuring its members operate to the highest ethical standard. RICS is incorporated by Royal Charter, which sets out the objects of the Institution and requires it to:

'maintain and promote the usefulness of the profession for the public advantage.'

Therefore, to practise as a member of RICS at any level, you must prove that you are ethically sound. Before introducing the RICS standards, it is worth considering what a professional is.

Key personal qualities of a professional include ability, confidence, responsibility, belief and respect, honour, reputation and trustworthiness.

A definition of professionalism was given by Howard Land FRICS in *Professional ethics and the Rules of Conduct of RICS* and states that:

> 'Professionalism is the giving of one's best to ensure that clients' interests are properly cared for, but in doing so the wider public interest is also recognised and respected.'

Professionals show high levels of ethical responsibility, ethics being defined as the 'science of morals'. As a branch of philosophy ethics is concerned with theories of goodness and badness or right and wrong and of morality.

In recent years additional focus has been placed on ethics and many have called for more openness and transparency in business dealings. Traditional business problems cannot always be resolved by legislation and experience, hence the need for professional institutions to develop their own ethical codes.

The RICS Rules of Conduct

Professional ethics are monitored by RICS Regulation, with reference to the RICS Rules of Conduct. The RICS Rules of Conduct for members are principles-based and are clear and simple rules that adopt the five principles of good regulation:

1 Proportionality

2 Accountability

3 Consistency

4 Targeting

5 Transparency

The rules positions RICS as a bold, cutting-edge professional regulator for the 21st century.

The Rules of Conduct for members cover those matters for which individual members are responsible and accountable in their professional lives. They apply equally to all members, wherever they are in the world and whatever their chosen field of activity. The RICS website (www.rics.org/rulesofconduct) holds a series of help sheets on different aspects of the rules.

- Members shall at all times act with integrity and avoid conflicts of interest and any actions or situations that are inconsistent with their professional obligations.

- Members shall carry out their professional work with due skill, care and diligence and with proper regard for the technical standards expected of them.

- Members shall carry out their professional work in a timely manner and with proper regard for standards of service and customer care expected of them.

- Members shall undertake and record appropriate lifelong learning and, on request, provide RICS with evidence that they have done so.

- Members shall ensure that their personal and professional finances are managed appropriately.

- Members shall submit in a timely manner such information, and in such form, as the Regulatory Board may reasonably require.

- Members shall co-operate fully with RICS staff and any person appointed by the Regulatory Board.

There are also Rules of Conduct for firms. These rules set out the standards of professional conduct and practice expected of firms registered for regulation by RICS.

- A Firm shall at all times act with integrity and avoid conflicts of interest and any actions or situations that are inconsistent with its professional obligations.

- A Firm shall carry out its professional work with due skill, care and diligence and with proper regard for the technical standards expected of it.

- A Firm shall carry out its professional work with expedition and with proper regard for standards of service and customer care expected of it.

- A Firm shall have in place the necessary procedures to ensure that all its staff are properly trained and competent to do their work.

- A Firm shall operate a complaints handling procedure. The complaints handling procedure must include a redress mechanism that is approved by the Regulatory Board.

- A Firm shall preserve the security of clients' money entrusted to its care in the course of its practice or business.

- A Firm shall ensure that all previous and current professional work is covered by adequate and appropriate professional indemnity insurance cover which meets standards approved by the Regulatory Board.

- A Firm shall promote its professional services only in a truthful and responsible manner.

- A Firm shall ensure that its finances are managed appropriately.

- A Firm which has a sole principal (i.e. a sole practitioner or a sole director in a corporate practice) shall have in place appropriate arrangements in the event of that sole principal's death or incapacity or other extended absences.

- A Firm registered for regulation must display on its business literature, in accordance with the Regulatory Board's published policy on designations, a designation to denote that it is regulated by RICS.

- A Firm shall submit in a timely manner such information about its activities, and in such form, as the Regulatory Board may reasonably require.

- A Firm shall co-operate fully with RICS staff and any person appointed by the Regulatory Board.

The candidate's knowledge

The knowledge and understanding required of an Associate candidate for this competency and for the online ethics module itself is wide-ranging. It covers the role, function and significance of RICS; an understanding of society's expectations of professional practice; RICS Rules of Conduct for both firms and members; and the general principles of law and the legal system, as applicable in your area of practice.

The RICS Rules of Conduct, policies and associated help sheets are available on the RICS website (www.rics.org/rulesofconduct). You should make sure that you are aware of the Rules for both members and firms.

To understand the role and function of RICS, you should carry out some reading around the structure of RICS, the various professional groups, group boards and committees, and the functions performed, such as advising the government on housing, taxation, planning and landlord and tenant issues and bringing an influence to bear on all relevant aspects of society. Apart from keeping abreast of developments in newspapers, a good source of information is the RICS magazine, *RICS Business*, and the various other weekly 'property' publications.

The other requirement involves you in understanding the role of the professional person and of society's expectations of such a person. The general principles of professional integrity (the 12 standards listed below) hold good in all professional-client situations. An important issue is that of ethics, which have been defined as a set of moral principles extending beyond a formal code of conduct.

RICS expects members to act both within the Rules of Conduct and to maintain professional and ethical standards. Members are expected to apply a set of 12 standards in order to meet the high standards of behaviour expected of them. These standards are to:

1. Act honourably – Never put your own gain above the welfare of your clients or others to whom you have a professional responsibility. Always consider the wider interests of society in your judgments.

2. Act with integrity – Be trustworthy in all that you do – never deliberately mislead, whether by withholding or distorting information.

3. Be open and transparent in your dealings – Share the full facts with your clients, making things as plain and intelligible as possible.

4. Be accountable for all your actions – Take full responsibility for your actions and don't blame others if things go wrong.

5. Know and act within your limitations – Be aware of the limits of your competence and don't be tempted to work beyond these. Never commit to more than you can deliver.

6. Be objective at all times – Give clear and appropriate advice. Never let sentiments or your own interests cloud your judgment.

7. Always treat others with respect – Never discriminate against others.

8. Set a good example – Remember that both your public and private behaviour could affect your own, RICS' and other members' reputations.

9. Have the courage to make a stand – Be prepared to act if you suspect a risk to safety or malpractice of any sort.

10 Comply with relevant laws and regulations – Avoid any action, illegal or litigious, that may bring the profession into disrepute.

11 Avoid conflicts of interest – Declare any potential conflicts of interest, personal or professional, to all relevant parties.

12 Respect confidentiality – Maintain the confidentiality of your clients' affairs. Never divulge information to others unless it is necessary.

More about these ethical standards, together with useful guidance, can be found on the RICS Regulation website (www.rics.org/regulation).

Conduct rules, ethics and professional practice

This mandatory competency is a huge and expansive topic and is tested by way of the online ethics module. However, you should also link relevant evidence and structured development to this competency.

You will need to keep this competency in perspective. The online ethics module will remain within the confines of the knowledge and experience that a person with limited experience in the surveying industry, and possibly occupying a fairly junior position, will have gained. Therefore you will not be expected to have first-hand experience of some of the more detailed areas of the Rules of Conduct, such as your firm's professional indemnity insurance or the members' accounts regulations but you will be expected to demonstrate your knowledge of the requirements of RICS in these areas and how this would be implemented in practice.

The ethics module may also test your knowledge around some of the basics, such as why RICS has Rules of Conduct, and aspects of the 12 standards.

Your training should incorporate a mixture of practical experience, structured reading and perhaps some CPD-type events on current issues. This competency is obviously also an ideal subject for your professional development.

You should develop your practical experience at work, with an emphasis on taking instructions, understanding and dealing with conflicts of interest, and applying the rules and the twelve standards.

You may wish to consider three main areas in reviewing your knowledge for this competency:

- *the role and function of RICS*: What have you read about RICS activities in recent months, what do you understand this role to be?
- *the 12 standards*: Can you think of a time when you have had to act within your limitations?; Can you think of occasions when you had to make a stand? How did you deal with this?
- *Rules of Conduct*: What do you understand about the Rules of Conduct, what do they require?

Some example issues to consider for this competency are:

- If you are successful with your Associate assessment and you decide to start your own business as a chartered surveyor what sort of things will you need to consider and do?
- How will you meet the requirements of the Rules of Conduct?
- Do you understand the need for insurances (employer's liability, public liability), a health and safety policy and equal opportunities policy?
- Will you use the RICS logo as set out by RICS and use only the appropriate alternative designation relevant to their professional group?

- What is client's money and how can a firm preserve the security of this?

- If you are approached in respect of an instruction and agree terms over the telephone what would you then need to do and what details would you need to provide?

- If you are successful in undertaking a project for a client and they send you a case of wine as a thank you, what should you do?

- You are working for a surveying practice and a friend asks you to ... [example of work from the relevant professional group] . . . as a favour as you are a surveyor, what should you do?

The RICS Regulation website (www.rics.org/regulation) provides an excellent source of information regarding the RICS Rules of Conduct.

Finally, you should of course, ensure that you demonstrate knowledge of this competency in your evidence and in your structured development – provide examples of structured reading and training and also of practical experience; make sure that all your detailed activities clearly comply with the Rules of Conduct and associated standards.

For further information and guidance on the RICS Rules of Conduct the DVD *RICS Rules of Conduct* available from RICS Books (www.ricsbooks.com) is a very useful resource. The DVD is designed to inform and challenge your knowledge and raise awareness of the depth and the breadth of the Rules of Conduct and what is expected of a professional. The book *Ethics and Professional Conduct for Surveyors*, also available from RICS Books, is also a useful reference point.

The online ethics module

In addition to submitting your evidence and structured development record, you must successfully complete the online ethics module before you can become an Associate.

Once you have submitted for Associate assessment, RICS will send you a personal link and password for the online ethics module. You will then have two weeks in which to complete this module. The module consists of brief ethical scenarios, each of which is followed by five possible solutions. In each case there is one ideal solution. You must select what you consider to be the ideal solution. There is then a final test consisting of 20 questions. You must pass this test before your Associate Assessment can be completed. The ethics module is based on RICS' 12 ethical standards.

If you do not pass the online ethics module you will be notified by RICS, and told when you can resit. Once you pass the ethics module RICS will accept that you have met the requirements for the mandatory competency.

There is a time limit of 12 months from the date you pass the module. If more than 12 months passes between that date and the date on which you pass your Associate assessment, you will need to retake the ethics module before you can be awarded the Associate qualification. RICS will expect you to maintain your ethical knowledge and understanding – passing the ethics module is not a 'once-and-for-all' achievement, but must be maintained throughout your career.

Figure 12 Example of the online ethics module

Some taster questions

To help you in your preparations for the online ethics module you might like to think about the following questions (the answers are given at the end of the page).

Professional ethics

Ethical standard: Act honourably/Comply with relevant laws and regulations

1. Your client has not yet had the time to ensure that his properties comply with new legislation. He instructs you to act in conflict with legal requirements by using delaying tactics to defer implementation. Your refusal to do so may harm your lucrative business relationship, leading to financial loss. However, to comply with the client's wishes may endanger the lives of his tenants. What should you do?

(a) The client is always right – I would act upon his instruction to maintain our professional relationship.

(b) Ignore the client's wishes – he should have known better than to ask me.

(c) Explain to the client that acting on his instruction would lead to illegal activity. If he still wishes to proceed then advise him that the contract with you will be terminated.

Ethical standard: Be accountable for all your actions

2. You misquoted for a job and find that you significantly underestimated the work involved. Half way through the job you have concerns that you may not satisfactorily be able to carry out the job for this fee. What should you do?

(a) Complete the undertaken job despite incurring a loss and recognising that you must be more careful in future.

(b) Back out of the job.

(c) Try to renegotiate with your client.

Ethical standard: Know and act within your limitations/Be open and transparent in your dealings

3. A client for whom you have worked on many jobs approaches you to undertake a project that will involve you in work in which you have no experience. You don't want to let the client down as they have been an excellent source of fees. What should you do?

(a) Take on the work and read up on the subject.

(b) Advise your client that this is not your area of expertise and recommend another surveyor who you know will do a good job.

(c) Refuse the commission with no explanation as you do not want your client to feel you are not able to do this job.

Ethical standard: Avoid conflicts of interest

4. You are selling land for a client/tendering a contract for a client and obtain an offer/tender bid from a company in which you are a major shareholder. What should you do?

(a) Sell the land/accept the tender if this is the best bid.

(b) Advise your client that you are unable to continue with the work.

(c) Advise all parties of your shareholding interest and your involvement with the client and seek written acceptance to continuing.

5. You are selling land/tendering a contract for a client and one of the developers promises you that if they buy the site/gain the contract they will instruct you to undertake work for them. What should you do?

(a) Encourage your client to sell/let the contract to this developer.

(b) Continue the sale/tender making no reference to this approach from the developer.

(c) Advise your client accordingly and seek their approval to you continuing with the sale/tender.

Ethical standard: Have the courage to make a stand

6. You are aware that properties that you manage/are contract managing the build for do not/will not comply with recently introduced fire safety legislation. New tenants are about to move in. Your client has asked you not to advise them of the discrepancy. What should you do?

(a) Let your client know that you are not prepared to endanger lives and explain that he must ensure the properties comply with the legislation.

(b) Accept the client's instruction as you are there to act on his behalf.

(c) Try to deter the tenants from moving in and encourage them to look for alternative accommodation.

Answers: 1a, 2a, 3b, 4c, 5c, 6a

Summary

- Conduct rules, ethics and professional practice is a mandatory competency for Associate assessment.
- There are many definitions of professionalism but importantly professionals have high ethical values and a strong sense of right and wrong.
- RICS has a responsibility to protect the public by ensuring its members operate to the highest ethical standard.
- RICS regulates members through Rules of Conduct, one for firms and one for members.
- RICS Rules of Conduct are based on 12 ethical standards.

- This competency is assessed primarily by way of the online ethics module but candidates should also link evidence and structured development to this competency.

9 The assessment

In this chapter we will take a look at the final Associate assessment which is, of course, facilitated through the Managed Learning Environment. The assessment is undertaken by two Associate assessors.

Let's firstly look at a quick checklist to make sure that everything is in order for your assessment.

Before submitting your portfolio

Checklist

Before submitting your portfolio for assessment, have you:

- undertaken the required number of years of work experience and/or gained the necessary qualification?
- submitted the required number of items of evidence for each competency and linked these to both the technical and mandatory competencies and has your Associate supporter verified that these are suitable items of evidence?
- checked that each item of evidence has been produced within the last four years?
- submitted a 300-word commentary with each item of evidence explaining how this relates to the technical competency and any mandatory competencies?

- proofread and spell checked all your evidence, structured development details and commentaries?
- checked that the evidence you have used gives the best examples of your work for that competency?
- recorded 48 hours of structured development over the last 12 months and recorded this in the Managed Learning Environment, including detailing the learning outcomes?
- sought your Associate supporter's confirmation that you have achieved the competencies?
- sought your Associate proposer's support that you are ready for assessment?
- uploaded declarations from your Associate supporter and Associate proposer?
- checked that everything is uploaded on the Managed Learning Environment (MLE)? When you click on 'I am ready to submit my evidence for Associate Assessment' the MLE will advise you of any missing documentation.

After submitting your portfolio

Checklist

After submitting your portfolio for assessment, have you:

- researched the RICS Regulation information and website?
- gained a good understanding of the RICS Rules of Conduct and associated guidance?
- revised information on professional ethics and practice?
- undertaken the online ethics module?

- passed the online ethics module?

The Associate assessors

The Associate assessors are specifically trained RICS members who assess your submitted evidence and structured development via the Managed Learning Environment (MLE) and decide whether you have met and satisfied the requirements of your chosen pathway. Two Associate assessors will review your evidence and structured development online; if you successfully meet the competency requirements and pass the online ethics module you will qualify as an associate.

Candidates who do not meet all the competency requirements will be referred, and given feedback on what additional evidence they need to submit (and possibly what extra experience they need to gain). The Associate assessors will base their decision on an all-round assessment, taking account of all your evidence, your commentaries and your structured development.

The assessors will in particular consider whether your evidence is:

- Acceptable – is there a suitable match between the evidence provided and the competence being claimed? Is the evidence valid and reliable?

- Sufficient – is there sufficient evidence to demonstrate fully the competence being claimed?

- Authentic – is the evidence clearly showing your own efforts and achievements?

- Current – does the evidence relate to current learning and experience (within the last four years)?

They will, of course, make their assessment against the detailed requirements of the competencies as set out in the Associate guidance.

Results

Approximately four weeks after you have been accepted for Associate assessment, RICS will notify you by email that your result is available online. It will be either 'Pass' or 'Refer'. You can access your result by clicking 'Progress' in the header bar and then selecting 'List assessments'.

If you pass the Associate assessment your membership will be upgraded from Associate candidate to associate (AssocRICS). You will be directed to the RICS members' zone to ensure all your details are correct. A welcome pack will be sent to you.

If you are referred you will be directed back to the MLE where your feedback report will be stored. The feedback report will display the competencies that require you to submit more evidence submitting and will include the assessors' comments. A new assessment has now been opened and you will be required to submit the requested amount of evidence. You do not have to resubmit everything, just the areas the assessor has highlighted – more evidence, structured development, work experience.

The Associate assessors will:

- provide feedback on each of the competencies;
- identify the pieces of evidence that satisfied them of your competence – these are 'banked' for a maximum of 12 months from the date of your result;
- provide feedback on your structured development; and
- give a clear explanation of what you will need to do in order to be ready for reassessment. For example, the feedback will:
    - state whether particular pieces of evidence did not meet the required standard or did not

clearly show your skills – in which case you will need to produce new or updated evidence for reassessment; and
- recommend specific experience you need to gain.

If additional experience is required, it will not be required to be undertaken for more than 12 months from the date of your result. This means that you will always be able to use any evidence you have 'banked' for at least one Associate assessment after a referral.

You should discuss the feedback with your Associate supporter and plan to resubmit within 12 months. If you go over that period, and there is more than 12 months between referral and resubmission, you will be starting again – that is, you will have lost the right to rely on the banked evidence, and all the evidence you submit must be new or updated in accordance with your feedback report.

You will be required to complete and record a minimum of four hours of structured development for each month between assessments. You can submit for reassessment as soon as you have assembled the new or updated evidence you need, and a minimum of four weeks has passed since your previous Associate assessment. Naturally, if the Associate assessors specify that you must complete a longer period of additional experience, you will not be able to resubmit for Associate assessment until you have completed that period.

Appeals

You have the right to appeal against a referral. However, you cannot appeal simply because you disagree with the decision of the Associate assessors. For an appeal to be successful you must be able to show fault in the way the Associate assessment was conducted, leading to an unfair decision; examples would be administrative error or procedural unfairness. You will have 21 working days from the date you received the result of your Associate assessment to make an appeal. If you wish to make an appeal you should contact RICS as there is a formal process to follow. An appeal fee will also be payable.

You should state, in no more than 1,000 words, the reasons for the appeal. No further supporting documentation is permitted and no representations may be submitted by another party, for example your Associate supporter or Associate proposer. Only an appeal directly from you (the candidate) will be considered, and no third party may appeal on your behalf.

The appeal will be considered by two appeal panel members who have experience of Associate assessment but were not involved in the original decision. If the panel dismisses the appeal, the referral will stand and you must provide the additional evidence specified in the feedback report before you can be reassessed.

If the panel allows the appeal, RICS will write to you advising you that the original Associate assessment result and feedback report are now void. You will be invited to reapply for Associate assessment with different Associate assessors using your existing evidence and structured development record. You may not submit any new documentation for the reassessment. The appeal fee will be refunded. If the two members of the appeal panel cannot reach a unanimous decision, your appeal will be allowed. The appeal panel's decision is final. There is no further right of appeal.

Quality assurance

RICS is committed to ensuring that the Associate qualification is supported by rigorous processes so that employers, clients and the public can have confidence that anyone who achieves it is competent to practise as an associate. RICS will audit all assessments through monitoring and comparing assessment outcomes and standards. This will not only help to ensure confidence in the qualification but also consistency in the assessment across pathways, countries and candidates.

RICS will select a number of Associate assessments for an audit as part of the quality assurance process. If your evidence is audited, you and your Associate supporter/proposer may be asked for further evidence that the work is all original and reflects your job role.

As part of the audit process, RICS may require you, after your Associate assessment, to participate in a verification interview. The interview is conducted by telephone by an RICS auditor. One in ten candidates will be subject to a telephone based interview. Its purpose is not to reassess your competence, but to verify the extent of your involvement in the work covered by your evidence, and the validity of the assessment. Any element of the Associate assessment is subject to audit. Associate assessors will nominate an Associate candidate for a verification interview if they have doubts about whether their evidence is genuinely original – for example, if they suspect plagiarism, or passing off another person's work as their own. The remainder will be selected randomly. If the auditor is not satisfied, the individual and employer concerned may be referred to RICS Professional Conduct for formal assessment.

Success

When you pass your Associate assessment you will then be able to use the letters AssocRICS after your name. I would urge you to use this on all your professional correspondence as it is an excellent way to show publicly your professionalism and your technical competence.

Progression from associate membership

After becoming an RICS associate, your professional development does not need to stop. The RICS Associate qualification is the first step on the journey to becoming a chartered surveyor.

In order to become chartered and to gain the letters MRICS you would need to undertake some further study and gain further experience.

Specific issues to consider when considering moving on to chartered membership are:

- Have your work activities progressed to an advanced level?

- Does your work involve giving reasoned advice?

- Have you gained, or are you able to gain a broader range of technical knowledge and skills than that required for Associate assessment?

- Have you gained, or are you able to gain, a higher level of technical competence than that required for Associate assessment?

- Have you (or will you be able to) achieve the additional mandatory competencies – Accounting principles and procedures and Business planning?

Generally, moving on from Associate to become a chartered surveyor will involve taking 900 study hours of the final level of an RICS-accredited degree-level qualification and gaining a further four years of experience. In effect, the study hours required equate to around six modules of a degree.

The study hours must usually meet the following requirements:

- map to the chosen pathway and competencies;

- total 900 study hours at final undergraduate or postgraduate level (generally this is 25% of a full undergraduate course, 75% of the final level of an undergraduate course or 50% of a postgraduate course – but see later regarding other options);
- widen and deepen knowledge and understanding of the technical competencies of the chosen pathway; be subject to assessment; be undertaken prior to the APC final assessment; and
- be approved by RICS.

Study provision will normally (at least initially) be from RICS partner universities, with the university offering the same course content and assessment methodology as for the full accredited course. RICS will require proof of successful completion of the 900 study hours, but will not require a specific exit award.

Individual modules normally comprise 150, 200 or 300 study hours; 900 study hours will therefore normally be six single modules.

Modules must map against a competency or part-competency and provide:

- knowledge and understanding plus application of that knowledge and understanding for competencies not included for AssocRICS; and
- deeper technical knowledge for competencies met for AssocRICS to ensure development beyond the AssocRICS level.

In order to widen access and opportunities for progression, up to 300 of the 900 study hours can be met by one of, or a mix of:

- work-based learning;
- independent study; and
- lifelong learning/in-house training.

However, these must be approved by RICS. No single component of less than 150 study hours will be considered.

During the minimum period of four years' post-AssocRICS relevant experience, you will need to further develop your levels of competence, building on the pathway competencies met for AssocRICS. After completing your study hours you will need to take the Assessment of Professional Competence (APC). Many of the competencies for the APC will be the same but taken to a higher level. Because a chartered surveyor is expected to have a broader range of knowledge and skills, most pathways will include additional technical competencies to the six taken for AssocRICS. Before APC final assessment you must have achieved all the technical competencies for your APC pathway.

The APC final assessment comprises of written submissions, some of which are prepared throughout your training period, and also a critical analysis. The critical analysis is a reflective critique of a project you have been involved with. In addition, you will attend a one hour interview with a panel of APC assessors to demonstrate your competence. Details of the APC can be obtained from the RICS website (www.rics.org/apc) or *isurv APC* (www.isurv.com/apc). You may also wish to purchase *APC: Your practical guide to success* and, for your employer, *Supervisors' and counsellors' guide to the APC*.

An alternative progression route is to complete an accredited degree. The Associate qualification is accepted for advanced entry to an accredited undergraduate degree. To achieve MRICS, levels two and three of an accredited undergraduate degree (2,400 study hours) must be completed. Studying part-time or by distance learning will normally take a minimum of three years, followed by one year's structured training after graduating and ending with the APC final assessment. In some circumstances, for example an associate with exceptional experience, entry to an accredited postgraduate award may be possible.

If you are interested in progressing through RICS you should speak to RICS and to RICS partner universities to explore your study options further. A full list of RICS partner universities (those offering accredited degrees) is available on the RICS website.

Fellowship

Attaining RICS fellowship (FRICS) is an exceptional mark of distinction. Normally only MRICS members with a minimum of five years service who are major achievers in their careers may apply to become a fellow of RICS. Fellows are leaders in the profession: active people who have completed unique projects or contributed to the profession. When the time comes the fellowship application process is designed to give you every opportunity to demonstrate your distinguished professional progression.

Summary

- Two assessors will assess your portfolio submission.
- The assessment will consider all the evidence, commentaries and structured development you have submitted.
- The assessors will take a comprehensive view of your competence.
- You will receive your results within four weeks from submission.
- If you are referred you will receive constructive feedback to help you to prepare for reassessment.

- You have the right of appeal against a referral decision if, in your opinion, there was fault in the way the assessment was conducted.
- RICS adopts a strict quality assurance procedure to ensure robust assessment.
- You may be asked to take part in a verification interview as part of the quality assurance process. This is not a reassessment.
- When you pass you will be able to use the letters AssocRICS.
- This need not be the end of your journey. AssocRICS is the first step on the journey to becoming a chartered surveyor.

10 Conclusion

As we have discussed throughout this book the Associate assessment is foremost a period of practical experience and the achievement of core personal and technical competencies. If you follow the guidance and ensure that you are learning the skills and reaching the levels of attainment required in the various competencies, you will be well on your way to a successful assessment. If the training and experience has been correctly put in place over the required period leading up to the assessment, the outcome should be a formality.

It is when candidates have not followed the various guidance that is available, and have not remained focused on the competency requirements, that the assessment results in an unsatisfactory outcome.

It is also essential that you do not forget about the mandatory competencies. These are every bit as important as the technical competencies. Ensure that you address these in your commentaries and that you link your evidence to these as well as to your technical competencies for your chosen pathway.

Vital points to bear in mind as you progress through the Associate assessment:

- You must have submitted the correct number of pieces of relevant evidence for each competency to the managed learning environment before being able to go forward for assessment. You should also ensure you have submitted commentaries with each item of evidence.

- The assessment is competency-based. Remember, you are being assessed to ensure you are 'competent to practise' overall as an RICS associate.

- Stay focused on the competency requirements for your pathway.

- Ensure you have completed at least the minimum period of work experience as confirmed to you by RICS before submitting evidence for assessment.

- Remember you can always change uploaded evidence up until the point you submit for assessment. You can use the Managed Learning Environment as a file for documents you may consider using for assessment.

- Use the official RICS Associate guidance and the MLE user guide to help you through the process.

- Don't be afraid to ask if you are unsure about anything. Your Associate supporter will be able to guide you in relation to work experience and structured development issues and the Associate support team at RICS will be able to answer any queries you may have regarding the process or the Managed Learning Environment.

- Throughout your Associate assessment remember that there is a wide range of help and support available for you and for your employer including:
    - your Associate supporter and proposer;
    - RICS regional training advisers;
    - the Associate support team at RICS;
    - the RICS website;
    - RICS Books;

Conclusion

- RICS Matrics;
- the RICS Library;
- *isurv Associate*;
- other Associate candidates – the *isurv Associate* channel and RICS Matrics provide opportunities for you to network with other candidates.

Remember that your learning and development never ends. Your professional competence will continue to be assessed by employers, clients and peer groups throughout your career. As the world around you changes in terms of consumer demands, law and technology, the content and focus of your learning and development will also change.

Therefore, when you become qualified, think of the RICS programme of continuous professional development (CPD) and lifelong learning as a vehicle to assist you with your essential and ongoing training and development needs, in a rapidly changing environment.

Finally, AssocRICS can either be the final destination of your journey or it can be a milestone on your way to becoming a chartered surveyor. You do not need to make the onward journey immediately but I would urge you to give serious consideration at some point in your career to taking the next step up. This really is a journey to success – 'gain the advantage' as the RICS Associate literature says!

11 Checklists

Assessment submission

Before submitting your portfolio for assessment you should check you have:

- Undertaken the required number of years of work experience and/or gained the necessary qualification;
- Submitted the required number of items of evidence for each competency and linked these to both the technical and mandatory competencies;
- Gained your Associate supporter confirmation that the evidence is suitable and demonstrates your competency;
- Only used evidence that has been produced within the last four years;
- Submitted a 300 word commentary with each item of evidence explaining how this relates to the technical competency and any mandatory competencies;
- Proofread and spell checked all your evidence, structured development details and commentaries;
- Ensured that you are happy that the evidence you have used gives the best examples of your work for that competency;
- Recorded 48 hours of structured development over the last 12 months and recorded this in the managed learning environment including detailing the learning outcomes;

- Sought your Associate supporters confirmation that you have achieved the competencies;
- Sought your Associate proposer support that you are ready for assessment;
- Uploaded declarations from your Associate supporter and Associate proposer;
- Uploaded on the MLE and clicking on 'I am ready to submit my evidence for Associate Assessment'. This will advise you of any missing documentation.

After submitting your portfolio you should check you have:

- Researched the RICS Regulation information and website;
- Gained a good understanding of the RICS codes of conduct and associated guidance;
- Revised information on professional ethics and practice;
- Undertaken the online ethics module; and
- Passed the online ethics module.

If you are referred you should ensure that you submit all additional evidence, structured development and work experience referred to in your feedback report.

Associate supporter action

Month 0

- Meet with Associate candidate
- Set out role of Associate supporter
- Agree dates for three monthly reviews
- Agree appropriate pathway and optional competency choices

- Prepare structured training plan

At three month intervals

- Meet with candidate
- Review progress against structured training plan
- Assess competencies and review evidence to support competency achievements
- Review and agree next stage of training and work experience
- Review structured development

At end of work experience period

- Confirm competency achievements or set extended training period
- Review structured development
- Review evidence placed in MLE
- Check all commentaries prepared by candidate
- Prepare declarations as supporter
- Arrange proposer declaration with candidate
- Agree revision plan for online ethics module

Ethics module

When preparing for the online ethics module you should ensure you are familiar with the following:

- What ethics means
- What professionalism means
- Principles of regulation
- Why the RICS has Rules of Conduct – who does this protect?
- RICS Rules of Conduct for members

- RICS Rules of Conduct for firms
- RICS 12 ethical standards
- Professional Indemnity Insurance – who needs it, what it is, how much you need, type of policy, levels of indemnity
- Client's money – what is it, how do you account for it?
- How to check the identity of your client
- How to comply with money laundering regulations
- What you would need to do if you set up in business on your own
- Lifelong learning – what is it, who must do it, what can it include, how much should you do?
- RICS logo and brand – how to use, regulations
- Registered firms – what this means, what is a registered firm, what a registered firm has to do
- Complaints handling procedures – what this should include, when to refer to it?
- Terms of engagement – written, what to include
- Gifts and inducements – should you accept, how to deal with them?
- Conflicts of interest – what a conflict of interest is, how to deal with one if it arises
- Quoting fees – rules against fee cutting
- Ethical dilemmas in your area of practice and how to deal with them

The 12 RICS ethical standards

1. **Act honourably** – Never put your own gain above the welfare of your clients or others to whom you have a professional responsibility. Always consider the wider interests of society in your judgements.

2. **Act with integrity** – Be trustworthy in all that you do – never deliberately mislead, whether by withholding or distorting information.

3. **Be open and transparent in your dealings** – Share the full facts with your clients, making things as plain and intelligible as possible.

4. **Be accountable for all your actions** – Take full responsibility for your actions and don't blame others if things go wrong.

5. **Know and act within your limitations** – Be aware of the limits of your competence and don't be tempted to work beyond these. Never commit to more than you can deliver.

6. **Be objective at all times** – Give clear and appropriate advice. Never let sentiments or your own interests cloud your judgement.

7. **Always treat others with respect** – Never discriminate against others.

8. **Set a good example** – Remember that both your public and private behaviour could affect your own, RICS' and other members' reputations.

The 12 RICS ethical standards

9 **Have the courage to make a stand** – Be prepared to act if you suspect a risk to safety or malpractice of any sort.

10 **Comply with relevant laws and regulations** – Avoid any action, illegal or litigious, that may bring the profession into disrepute.

11 **Avoid conflicts of interest** – Declare any potential conflicts of interest, personal or professional, to all relevant parties.

12 **Respect confidentiality** – Maintain the confidentiality of your clients' affairs. Never divulge information to others unless it is necessary.

The RICS Rules of Conduct for members

- Members shall at all times act with integrity and avoid conflicts of interest and any actions or situations that are inconsistent with their professional obligations.

- Members shall carry out their professional work with due skill, care and diligence and with proper regard for the technical standards expected of them.

- Members shall carry out their professional work in a timely manner and with proper regard for standards of service and customer care expected of them.

- Members shall undertake and record appropriate lifelong learning and, on request, provide RICS with evidence that they have done so.

- Members shall ensure that their personal and professional finances are managed appropriately.

- Members shall submit in a timely manner such information, and in such form, as the Regulatory Board may reasonably require.

- Members shall cooperate fully with RICS staff and any person appointed by the Regulatory Board.

The RICS Rules of Conduct for firms

- A Firm shall at all times act with integrity and avoid conflicts of interest and any actions or situations that are inconsistent with its professional obligations.

- A Firm shall carry out its professional work with due skill, care and diligence and with proper regard for the technical standards expected of it.

- A Firm shall carry out its professional work with expedition and with proper regard for standards of service and customer care expected of it.

- A Firm shall have in place the necessary procedures to ensure that all its staff are properly trained and competent to do their work.

- A Firm shall operate a complaints handling procedure. The complaints handling procedure must include a redress mechanism that is approved by the Regulatory Board.

- A Firm shall preserve the security of clients' money entrusted to its care in the course of its practice or business.

- A Firm shall ensure that all previous and current professional work is covered by adequate and appropriate professional indemnity insurance cover which meets standards approved by the Regulatory Board.

- A Firm shall promote its professional services only in a truthful and responsible manner.
- A Firm shall ensure that its finances are managed appropriately.
- A Firm which has a sole principal (ie a sole practitioner or a sole director in a corporate practice) shall have in place appropriate arrangements in the event of that sole principal's death or incapacity or other extended absences.
- A Firm registered for regulation must display on its business literature, in accordance with the Regulatory Board's published policy on designations, a designation to denote that it is regulated by RICS.
- A Firm shall submit in a timely manner such information about its activities, and in such form, as the Regulatory Board may reasonably require.
- A Firm shall cooperate fully with RICS staff and any person appointed by the Regulatory Board.

FAQs

Q: What are the enrolment dates for the Associate qualification?

A: Providing your pathway is available, you can enrol on the Associate qualification at any time; however your eligibility to apply for Associate Assessment is dependent on attainment of relevant work experience and/or vocational qualifications. The relevant work experience can include experience gained prior to Associate enrolment. You will need to demonstrate that you meet all the competency requirements before applying for assessment.

Q: What are the requirements for the Associate assessment?

A: The Associate qualification has been developed to recognise the experience, qualifications and skills of an applicant working within the land, property and construction sectors. The key requirements are for you to have a minimum of four years' relevant work experience, which can be retrospective and to provide evidence against each of the technical competencies to prove that you have reached the level of competence RICS is looking for in an Associate. The period of required experience can be reduced if you hold professional, academic or vocational qualifications at the appropriate level.

Some qualifications will grant you direct entry to Associate. As a direct entry candidate you will not be required to submit evidence against competencies, however you will still be required to successfully pass the ethics test.

You will also be required to provide details of 48 hours of structured development that you have completed in the last 12 months and successfully complete the RICS online ethics test.

Q: What does RICS mean by competency?

A: A competency is a specific task or function that you must prove that you are able to perform to an agreed standard to become an Associate. There are two distinct types of competency that you will be required to meet.

Technical competencies – These are the primary skills of your chosen pathway. You will be asked to demonstrate that you are competent in these skills by submitting work-based evidence against each competency.

Mandatory competencies – These are the skills that are common across all pathways and are deemed essential to becoming an Associate. They are:

- Conduct rules, ethics and professional practice
- Client care
- Communication and negotiation
- Health and safety
- Sustainability
- Teamworking
- Data management
- Conflict avoidance, management and dispute resolution.

You will need to demonstrate these competencies through a mix of your structured development records and your technical competency submissions.

FAQs

Q: How many competencies will I need to meet?

A: You will be required to submit evidence against around six technical competencies (there is some variance between pathways), and eight mandatory competencies.

Q: What does RICS mean by structured development?

A: Structured development is training or learning that enables you to gain extra skills and knowledge that can be applied in your day to day role. This structured development could be in the form of distance learning, formal training courses or structured reading. It must be planned and recorded with all learning outcomes evaluated.

When you apply for your Associate assessment we will ask you to provide records of a minimum of 48 hours structured development gained within the 12 months prior to your assessment.

Q: How do I access the Managed Learning Environment (MLE)?

A: Once you have registered on the Associate qualification you will be emailed your RICS membership number. This will give you access to the member zone on rics.org where you will be able to set your password. Once this is complete you will be able to access the Managed Learning Environment (MLE) directly via; mle.rics.org.

Q: What do you mean by evidence?

A: Evidence is actual documented examples of your work which show that you meet specific competency requirements. For each pathway, candidates will be advised of the types of documented evidence to be presented.

Q: What can I upload as evidence?

A: Each pathway will have different requirements. You can upload evidence from the last four years of your career, however, it is important that at least one piece of evidence for each competency is from the 12 months prior to your Associate assessment. However, Associate candidates with at least 10 years experience who are now working in a narrower but more senior capacity would not be required to submit one piece of evidence from the last 12 months for up to three technical competencies but instead could use 'older' evidence bringing it up to date in the commentary, possibly with reference to work the candidate has supervised.

Q: Can I upload more than the minimum amount of evidence?

A: You can use the MLE to record experience and structured development throughout the time you are training. When you prepare for assessment you will be required to indicate on the MLE which pieces of evidence you would like to be assessed. You will only be able to submit the required number of pieces of evidence and structured development for assessment. We recommend that you seek advice from your supporter regarding suitability of evidence and structured development. You can use the MLE as a file repository until such time as you submit your evidence for assessment.

Q: What is the purpose of the commentary?

A: The 300 word commentary required for each item of evidence submitted allows you to demonstrate how the evidence has helped you to meet that competency and any relevant mandatory competencies. The commentary should fully explain your role in the work submitted.

Q: Can I use evidence I have used for an NVQ towards my Associate assessment?

A: Yes, the vocational experience required for relevant NVQs should be closely related to the Associate competencies and it is likely, therefore, that the same evidence can be used. This must be uploaded to the MLE in the usual way and supported with a 300 word commentary. Evidence from NVQs can include Storyboards.

Q: Who signs off competencies/submissions and what support do I need to undertake the assessment?

A: When you put forward your portfolio of evidence for assessment you will need to have it signed off by somebody who is familiar with your work and able to verify that it is your own and meets the requirements for Associate assessment, as defined by RICS, for example your line manager.

You also need to have your membership application proposed by an existing member of RICS (FRICS, MRICS and AssocRICS).

It is considered best practice to have a supporter within your firm who can provide you with support and guidance through your training period. Typically this would be your line manager.

Q: If the Associate assessment is completely online how will RICS be able to ensure consistency of assessment?

A: RICS has developed a rigorous auditing process which will sample a percentage of the assessments carried out to ensure the candidates being assessed are meeting the competencies, the standards set are being adhered to and the assessors are assessing at a consistent level. This monitoring of the Associate assessment will also enable RICS to keep the qualification relevant to the market requirements.

Q: What happens if I fail the online ethics test?

A: If you fail the online ethics test you can take the test again but you must wait at least 24 hours before you can do this.

Q: What happens if I pass the assessment but not the ethics test?

A: To become an Associate you will need to pass both the ethics test and the assessment. If you fail the ethics test you will be able to take the test again, however, you will need to wait 24 hours before doing so.

Q: Will I have to attend an interview for Associate assessment?

A: No, there is no interview for the Associate assessment.

Q: What is a corporate partner of RICS?

A: For the Associate assessment RICS has set up a corporate partner scheme which is available to firms that have five or more candidates on the Associate qualification. Firms are required to support the assessment by providing a minimum of two assessors and benefit from corporate invoicing and support services.

Q: What happens when I get my result?

A: Once your assessment is complete, approximately four weeks after you have submitted for assessment, you will be emailed by RICS to tell you that your result is available to view on the Managed Learning Environment (MLE). If you pass the assessment your membership will be upgraded from Associate candidate to Associate – AssocRICS. You will receive your membership welcome pack in the post which will include your membership card and diploma.

Q: What happens if I am unsuccessful at the assessment?

A: If you have been unsuccessful you will receive an email directing you to your feedback report on the MLE. The assessment panel will provide you with feedback on each of the competencies and your structured development and will detail what you need to do to be ready for reassessment including how to pay the reassessment fee.

Q: Can I progress to chartered membership after becoming an Associate?

A: The progression from Associate membership (AssocRICS) to Chartered membership (MRICS) is a combination of further experience and academic study. Associate members will be required to complete a further 900 study hours (equivalent of 6 modules) from an RICS accredited degree, and have four years post qualification experience as an Associate member before embarking on the Assessment of Professional Competence (APC). As an Associate you will be able to gain advanced entry to an accredited degree. You can see what courses are available by visiting www.ricscourses.org

If you already meet the requirements of an APC Graduate Route to membership you would not be required to complete the further study hours and could enrol onto the APC as soon as you are ready.

Q: Where could I take the 900 study hours from Associate to chartered status?

A: This study option is mainly offered by RICS partner Universities (ie those offering RICS accredited degrees). However, there is also the option for other Colleges and Universities to apply for approval of modules to offer this option. Additionally, some of the study may be offered by employers. Any course of study must be approved by RICS.

Q: If I already had a lot of experience before becoming an Associate Member would I still need to have four years post qualification experience when progressing on?

A: Yes, this applies to all candidates.

Q: Can the 900 study hours be reduced?

A: No, the 900 study hours applies to all candidates in order to ensure consistency.

Q: If I take the Associate route will I take the APC final assessment?

A: Yes, the APC final assessment will be the same as for all APC candidates and will include written submissions and attendance at an interview.

Q: Which pathways are available for me to take the Associate assessment?

A: The pathways available at the time of writing are:

- Quantity surveying and construction
- Project management
- Facilities management
- Residential surveying and valuation
- Residential estate agency
- Building surveying
- Building control
- Commercial property management
- Land engineering surveying
- Hydrographic surveying
- Residential property management
- Valuation

Index

appeals
  Associate assessment
               157–158, 164
apprenticeships   29–32
  levels
    advanced
      apprenticeships
      (equivalent to two
      A-level passes)   30
    apprenticeships
      (equivalent to five
      good GCSE passes)
               30
    higher apprenticeships
               30
arbitration   46
assessment plan
  step to achieving
    qualification   6
assessment ready
    candidates   9–10
  required experience   9
Associate assessment   14, 32, 153, 163–164, 165
  after submitting portfolio
    appeals   157–158, 164

Associate assessors   155, 163
  checklist   154
  quality assurance
               158–159
  results   156–157, 163
  success   159
before submitting
  portfolio, checklist
        153–154, 168–169
competency-based
  assessment *see*
  competency-based
  assessment
gaining work experience
  see structured
  training
NVQ link   27–29
progression, vital points
  to consider   165–167
progression from
  associated
  membership   160–163, 164, 167
  fellowship   163
readiness   135–137

step to achieving
 qualification 7
structured development
 see structured
 development
Associate assessors 12, 155
Associate competencies 33
 mandatory competencies
  33–35
 technical competencies
  35–39
Associate pathway 15, 32
 mandatory competencies
  see mandatory
  competencies
 RICS practice statements
  and guidance 38
Associate proposer 12, 13
Associate supporter 11–12, 13
 action checklist 169–170
 role in structured training
  86–89, 104
 three-monthly reviews
  90–91, 98–103, 104
AssocRICS *see* RICS
 Associate
 (AssocRICS)

candidates 8, 13
 assessment ready 9–10
  required experience 9
 direct entry 8–9
 enrolment ready 10–11
 other candidates 13
career development 2–3
client care competency
  39–40

communication and
 negotiation
 competency 41–43
competency-based
 assessment 15–19, 32
 a competency 16–19
 competence 19
 core competencies 15, 32
 mandatory competencies
  15–16, 32
 National Vocational
  Qualifications *see*
  National Vocational
  Qualifications
  (NVQs)
 optional competencies
  15, 32
conduct rules, ethics and
 professional practice
 see professional
 ethics
conflict avoidance,
 management and
 dispute resolution
 procedures
 arbitration 46
 competency 44–47
 independent expert 46–47
 mediation 46
 Professional Arbitration
  on Court Terms
  (PACT) 45–46
continuous professional
 development (CPD)
  4, 167
core competencies *see*
 technical
 competencies

Index

data management
   competency   47–54
   Data Protection Act 1998   49–51
   Environmental Information Regulations   52–53
   Freedom of Information Act 2000   51–52
   Freedom of Information (Scotland) Act 2002   52
   issues for consideration   53–54
direct entry candidates   8–9
dispute resolution procedures *see* conflict avoidance, management and dispute resolution procedures

enrolment ready candidates   10–11
ethics module *see* professional ethics
evidence of competence
   submitting evidence   74–76
      choosing evidence   76–77
      linking evidence to competencies   78–84
      uploading evidence   77–78
      using existing experience   73–74

fellowship
   attaining RICS fellowship (FRICS)   163
frequently asked questions   177–184

health and safety
   competency   54–59
   personal property of others   58–59
   personal safety   56–57
   relevant legislation   55–56
   risk assessments   56
   safety of property   57–58

independent expert   46–47

Managed Learning Environment (MLE)   7, 125, 137–138
   adding evidence   77, 129–130, 169
   confirmation of submission   130
   homepage   127–129
   selecting competencies   129
   managed learning environments generally   125–127
   ready for assessment   135–137
   structured development screens   120, 130–134
portfolio
   assembly of portfolio of work-based evidence   7

viewing 134–135
mandatory competencies
    15–16, 32, 33–35,
    71–72, 165
  client care 39–40
  communication and
    negotiation 41–43
  conduct rules, ethics and
    professional practice
    44
  conflict avoidance,
    management and
    dispute resolution
    procedures 44–47
  data management 47–54
  evidence *see* evidence of
    competence
  health and safety 54–59
  sustainability 59–61
  teamworking 62–70
mediation 46
National Vocational
    Qualifications
    (NVQs) 21–22, 32
  achieving 22–23
  benefits 23–24
  bodies responsible for
    setting standards
    25–27
  definition of levels 24–25
  link to Associate
    assessment 27–29
  Qualifications and
    Curriculum
    Framework (QCF)
    26–27
  Scottish Qualifications
    Authority (SQA) 27

negotiation *see*
    communication and
    negotiation
    competency

online ethics module *see*
    professional ethics
online qualification process
    4
optional competencies *see*
    technical
    competencies

personal property
  property of others 58–59
  safety 57–58
personal safety 56–57
portfolio
  after submitting
    appeals 157–158
    Associate assessors 155
    checklist 154
    quality assurance
    158–159
    results 156–157
    success 159
  before submitting,
    checklist 153–154,
    168–169
  Managed Learning
    Environment (MLE)
    assembly of portfolio of
    work-based evidence
    7
  viewing portfolio
    134–135

Index

Professional Arbitration on Court Terms (PACT) 45–46
professional development *see* structured development
professional ethics 139, 151–152
  being professional 139–140, 151
  RICS Rules of Conduct 140–143, 151
  candidate's knowledge 143–147
  conduct rules, ethics and professional practice competency 44, 145–147, 151
  online ethics module 148–151, 152
    checklist 170–171
    competency 44
    step to achieving qualification 7
  RICS 12 ethical standards 172–173

qualification 1–2
  benefits
    career development 2–3
    client confidence and employer assurance 3–4
    flexible online qualification process 4
    professional knowledge/information 4
    RICS Matrics 4
    status and recognition 3
  steps to achieving
    assessment plan 6
    Associate assessment 7, 14
    ethics module 7
    portfolio of work-based evidence 7
  qualification
  steps to achieving registration 6
Qualifications and Credit Framework (QCF) 19–21
quality assurance
  Associate assessment 158–159

records
  structured development
    access through MLE homepage 130–134
    completion 120–123
regional training advisers (RTAs) 12–13
registration
  step to achieving qualification 6
  structured training 92–98
  gaining work experience 89–90
results
  Associate assessment 156–157, 163
RICS Associate (AssocRICS) 1, 13–14

189

areas of work 5
Associate assessment
   success 159, 164
Associate assessors 12
Associate proposer 12, 13
Associate supporter 11–12, 13
candidates 8, 13
  assessment ready 9–10
  direct entry 8–9
  enrolment ready 10–11
  other candidates 13
  progression from associate membership 160–163, 164, 167
  fellowship 163
  qualification *see* qualification
  regional training advisers (RTAs) 12–13
  routes to membership 6–7, 13
RICS Associate Managed Learning Environment (MLE) *see* Managed Learning Environment (MLE)
RICS fellowship (FRICS)
  attaining 163
RICS Matrics 4
RICS membership
  route to various grades 2
RICS Rules of Conduct 44, 140–143, 145–147, 151
  firms 175–176
  members 174

risk assessments 56

safety of property 57–58
Scottish Vocational Qualification (SVQ)
Scottish Qualifications Authority (SQA) 27
structured development 107–108, 124
  professional development, concept 105–107
  record
    access through MLE homepage 130–134
    completion 120–123
  stages
    appraisal 108–109
    development 111–113
    planning 110–111
    reflection 113–115
  types of activities 115
    organised learning 118
    personal 115–118
    work-based learning 118–120
structured training 86, 104
  halfway point actions 103
  registration 92–98
  three-monthly reviews 98–103, 104
  work experience 86, 103
    Associate supporter's role 86–89, 104
    completing minimum experience required 91

Index

progressive assessment 91–92
registration 89–90
three-monthly reviews 90–91
sustainability competency 59–61

teamworking
  benefits 64–65
  competency 62–70
  team
    effective, characteristics 66
    leadership 68–69
    meaning 63
    reward 70
    role theories and selection 66–68
    size 68
    stages of development 65–66
    training and learning 69–70
    types 63–64
technical competencies 35–39, 71, 165
  core competencies 15, 32, 35
  evidence *see* evidence of competence
  example 36–38
  knowledge development 38–39
  optional competencies 15, 32, 35–36
three-monthly reviews
  structured training 98–103, 104
  gaining work experience 90–91
training *see* structured training

vocational qualifications 19
  apprenticeships *see* apprenticeships
  National Vocational Qualifications *see* National Vocational Qualifications (NVQs)
  Qualifications and Credit Framework (QCF) 19–21

work experience *see* structured training

RICS contact details and further information

There are numerous ways of contacing RICS and a wealth of information sources online for APC candidates.

RICS Contact Centre

T: +44 (0)870 333 1600

F: +44 (0)20 7334 3811

E: apc@rics.org

RICS website

The RICS website provides details on the routes to membership, how to enrol and how to apply for the final assessment. You can also download the APC guidance, templates, Excel workbooks and pathway guidance and find your nearest regional training adviser and APC doctor. Visit www.rics.org/apc

RICS Books

T: +44 (0)870 333 1600

F: +44 (0)20 7334 3851

E: mailorder@rics.org

www.ricsbooks.com

isurv

isurv is an online knowledge resource for property professionals, brought to you by RICS. The *isurv* 'channels' cover a huge range of surveying topics, mixing expert commentary with official RICS guidance. Visit www.isurv.com

The *isurv* APC channel also holds excellent support material and provides links to information relevant to achieving each competency. Visit www.isurv.com/apc